ANIMAL SCIENCE, ISSUES AND PROFESSIONS

NEOTROPICAL AND CARIBBEAN AQUATIC MAMMALS

PERSPECTIVES FROM ARCHAEOLOGY AND CONSERVATION BIOLOGY

Animal Science, Issues and Professions

Additional books in this series can be found on Nova's website under the Series tab.

Additional e-books in this series can be found on Nova's website under the e-book tab.

ANIMAL SCIENCE, ISSUES AND PROFESSIONS

NEOTROPICAL AND CARIBBEAN AQUATIC MAMMALS

PERSPECTIVES FROM ARCHAEOLOGY AND CONSERVATION BIOLOGY

A. SEBASTIÁN MUÑOZ
CHRISTOPHER MARKUS GÖTZ
AND
ELIZABETH RAMOS ROCA
EDITORS

nova publishers
New York

For permission to use material from this book please contact us:
Telephone 631-231-7269; Fax 631-231-8175
Web Site: http://www.novapublishers.com

Additional color graphics may be available in the e-book version of this book.

Library of Congress Cataloging-in-Publication Data

ISBN: 978-1-63321-306-7

Published by Nova Science Publishers, Inc. † New York

CONTENTS

INTRODUCTION

A. Sebastián Muñoz, Christopher Markus Götz and Elizabeth Ramos Roca

OBJECTIVES OF THE BOOK

This book aims towards a comprehensive presentation of the use of aquatic mammals by selected past human populations of the recent and remote times of the American continent. Focusing primarily on zooarchaeological data, but including also historical and biological information, we intend to create a rounded image of the importance those heavy animals had for the ancient peoples of the Americas, in spite of the difficulties that sometimes, but not always, their hunting and catching implied. Until the present, we lack a clear picture, at a continental scale, of the importance that these animals had for human consumption and other purposes, such the use of their remains as raw material, for rituals, among others, from a long term perspective. In consequence, in order to evaluate the diversity and changes in the way humans interact with these animals throughout time, we need first to review what the kinds of evidences we have and how and what these contrast in different regions of the continent. Analyze that kind of differences is precisely one of the main objectives of this book. The zooarchaeological tendency of the book relates to the specialization of the three editors and the majority of the authors; many are experts in the field of zooarchaeology, working at different areas in the Neotropical environments of the American continent.

THE AQUATIC MAMMALS TREATED IN THIS BOOK

Aquatic mammals are among the most endangered taxa throughout the world, due to industrial and domestic contamination and unmeasured resource use by the modern human societies (Trujillo et al. 2013, National Geographic, 1996). To investigate on the natural history of these animals and its relationships with humans along time in the Neotropics is of most importance, since this biogeographic region is dominated by oceans. This feature, in turn, conditions the nature of the prevailing faunas, which, for instance, in the Southern tip of South America are dominated by marine mammals (Redford & Eisenberg 1992).

Accounts about prehistoric times often idealize the role of aboriginal groups, emphasizing the supposedly wise use of nature, especially by American Indians. In an attempt to review critically these interpretations, we present, among other data, relevant information in this book that describe precisely the impact that also the Indian populations caused on aquatic mammal populations. The presentation of those data are aiming to increase our understanding of the relationship between humans and nature, and we hope that a critical assessment can help to contribute to the actual discussions about the use and conservation of sea mammals and create decisions towards the protection of those fascinating taxa, especially considering their relation to human cultures throughout the Holocene, as it is evidenced in the majority of the chapters in this book.

The aquatic mammal species we refer to in this publication share all one common characteristic, as they spend their complete or near complete life cycle in fresh or salt water. Yet, the defining character of the taxa treated in this book is not only based on the life cycle alone, but also on their occurrence in the archaeological record; remains of sea otters, beavers or capybaras are not frequently found in the archaeological record, although those small aquatic mammals might also have been used by the aboriginal peoples of the Neotropics. Big aquatic mammals, as the ones we primarily focus on in this book, are known, instead, from numerous archaeological excavations throughout the continent, and the examples discussed in the 8 chapters give a vivid example of their importance for the pre-Hispanic peoples of the Neotropics.

The taxa we discuss in the following chapters come from three orders, Sirenia, Cetacea and Carnivora. Out of the Sirenia (sirenids), the only species treated in this book is the manatee (*Trichechus manatus*). The chapters from Ramos and Pérez, and Götz, Valadez and Rodríguez discuss the importance of this animal among the pre-Hispanic Indians of the tropical and subtropical regions of the Neotropics. The carnivora (carnivore mammals) comprise in this publication only seals and their relatives, as well as otters. While archaeological remains of otters (here *Enhydra lutris*) are limited, as to the present state of knowledge, to contexts in Baja California (Mexico), described in the chapter of by Guía-Ramírez, sea lions (*Zalophus californianus* and *Otaria flavescens*) are treated in chapters that deal with human exploitation of oceanic resources in both, the northern (again, see chapter Guía-Ramírez) and southern parts (see chapter Zangrando, see chapter Borella), of the Neotropics. Mentions of seals are also distributed in chapters about the northern and southern part of the Neotropics, as the South American fur seal (*Arctocephalus australis*) and the South American sea lion (*Otaria flavescens*) are included in the discussions of the chapters by Borella and Zangrando, while the role of human exploitation of the West Indian seal (*Monachus tropicalis*) is analyzed in the chapters by Götz, Valadez and Rodríguez and Ramos Roca and Pérez Iglesias.

The cetacea, whales and dolphins, show, in comparison with the other taxa, a somehow less involvement with human societies, probably because their acquisition was more difficult and animals could not be hunted on shore or in shallow waters. Nevertheless, there are several examples of cetacean remains found in association with cultural remains, mostly isolated finds that were made at archaeological sites. In this sense, it is especially the chapter of Corona, Carmona and Arroyo that presents such a case from Ancient Mexico, where a whale rib (from a *Balaenoptera* sp) was found in a religious offering at an ancient site of Oaxaca (Mexico), while Götz, Valadez and Rodríguez present other examples of cetacean specimens (mostly vertebras) from archaeological sites of southeastern Mesoamerica.

The prehistoric information on aquatic mammals derived from archaeological data illustrated in the previously mentioned chapters is complemented in this book with two additional perspectives. On the one hand, we have Carvajal´s review of the use of aquatic mammals by human populations in Northern South America from ethnographic and ethnohistoric accounts from the 16th to 19th century. On the other, from a conservation biology point of view, Túnez and Pimper provide both a review on "the recorded presence and conservation status of marine mammals in southern South America" and a discussion "about biology, population numbers and trends, mayor threats, conservation genetics and population structure" of some of the most studied southern South American species of marine mammals in current times.

The Neotropics

The Neotropics are one of the most diverse regions on Earth. Among other ecosystems, this biogeographic region includes more tropical rainforest than any other in the world, and some of the driest deserts as well. As Muñoz and Mondini (2008) point out, this diversity, which has many implications on faunal communities, impinges directly on human-animal interactions and the resulting fossil record. In this book, we define the Neotropical region *sensu lato* (after Morrone, 2013) that is including both the Mexican and South American transitions zones. As this author highlights, the referred scheme has been largely used after Sclater´s proposal and Wallace´s inclusion of the concept in his *Geographical Distribution of Animals* masterpiece, from 1876. Regarding this book, the scheme is particularly useful for the topics reviewed here, since biogeographers researching on vertebrates mostly agree with it (Morrone 2013, see also Cox 2001).

Geographically, the Neotropics comprise a vast territory from Northern Mexico to the Southern tip of South America, that can be divided in several regions, namely: Mexico and Central America; the Antillean region, including the West Indies; the Brazilian region, which refers to tropical South America, and the Chilean region, which includes temperate South America (Wallace 1876, in Morrone 2013; see also Cabrera & Willink 1980). The internal variability displayed in the Neotropics is remarkable, and is expressed in the faunal communities that characterized this region, such as population densities, body sizes, etc. The particular history of the region has impinged on this configuration, as the consequence of the Pliocene faunal interchange between North and South America and the successive climate changes clearly illustrate. For instance, temperate South America has currently many ecological niches only partially occupied by mammals, as compared to its Northern counterpart. Hence the current South American ungulate fauna is one of the best examples of unsaturation, and it is markedly deficient as compared to that prior to the late Pleistocene extinctions (see Redford & Eisenberg 1992).

Ecological conditions, such as those referred above, are of importance when considering human insertion in the Neotropical faunal community, since humans would have had few competitors when peopling the continent. The former would have certainly affected the way humans included aquatic mammals in their diet from that time on. For instance, human exploitation of pinnipeds in Southern Patagonia during Holocene would have not been limited by other terrestrial competitors on land, unlike other regions of the world, where carnivores compete for those resources as well (e.g. South Africa, see Klein et al. 1999), which is certainly the case for other neotropical subregions.

The physical configuration of the Neotropics has also to be considered when human use of aquatic mammals is analyzed. As Morello (1984) point out, an important part of the Neotropics -South America- is a peninsula within the oceanic hemisphere. Hence, continentality does not increase with latitude in this region being oceanity, instead, the climatic feature that follows that latitudinal gradient. Other structural features are important as well, such as the low structural diversity of terrestrial habitats (Morello 1984). Following this author, we have to highlight the importance of the Orinoco and Amazon basins, which constitute the most extensive hyperhumid areas of the world. This certainly is of importance when human insertion in neotropical ecological contexts is considered, since it offers a huge amount of energy in the form of aquatic resources.

Cultural Diversity of the Regions Treated in This Book

Having into account the importance of the Neotropics, we define in the next paragraphs the cultural diversity that comprises the different regions, beginning with Mexico and Central America, followed by northern South America and the Antilles and then by the southern part of the continent.

In what refers to the northern and central parts of the area treated, (Mexico and Central America) the cultures whose interaction with aquatic mammals and especially sea mammals is scrutinized in this volume developed on a similar chronological scale. The concept of 'Mesoamerica', treated in the chapters of Corona et al., and Götz et al., was formally defined by Paul Kirchhoff in 1943 (1967; see Matos 1994:51-53), based on similarities between the remains of high civilizations that encompassed the Pre-Hispanic population and its material evidence between the southeastern USA and the Isthmus between El Salvador and Honduras (Haberland 1991:128; Prem 1989:3). At its northern limits, the other vast region culturally defined, 'Aridamerica', includes not only the northern part of Mexico and the southern USA, but also the peninsula of Baja California (see Nárez 2000:123-127), an area described in the chapters by Guía-Ramírez. In late Prehispanic times, Mesoamerica represented an extraordinary cultural and environmental diversity, with at least 14 subareas (Adams 1991:14-15; Smith and Masson

2000:2). Mesoamerica is considered one of the main centers of early plant domestication in the Americas (Niederberger 1996:11), and is particularly important for early domestication of the agricultural complex of maize, beans and squash that was shared by all the cultures of the area. Equally important were the domestication of turkey and honey-bees, as well as the pan-Mesoamerican use and breeding of the dog, and the probable husbandry of white-tailed deer and other forest species (Masson & Peraza 2008; Valadez 2003).

The history of Mesoamerica started with the Paleoindian period (approx. 13,500 – 10,000 BP, according to Lohse et al. 2006, among others), defined by the lithic Clovis points (for the north of Mesoamerica) and Fishtail points (for the southern area). In this first occupation of Aridamerica and Mesoamerica, simple bands of hunter-gatherers specialized in the hunting of megafaunal mammal groups of the final Pleistocene (Lohse et al. 2006; Mirambell 1994, 2001; Nárez 2000:126). The Mesoamericans may have first become sedentary during the Archaic period (approx. 10,000 – 2,900 BP, according to Lohse et al. 2006, among others)[1], when climate change and other factors reduced the availability of megafauna, but started to offer ecosystems with multiple sources for exploitation (Garcia Bárcena 2001). The first indications of domestication of one of the emblematic plants of Mesoamerica, maize, also appear around this time (9,100 BP from genetic data indicating domestication in the Rio Balsas region of lowland tropical Mexico, 7,100 BP from maize phytoliths in coastal Tabasco, and 6,200 BP from macrobotanical remains in highland Mexico, see Pohl et al. 2007). In Aridamerica, instead, maize appeared not before 2,500 BP (Nárez 2000:127). The first evidences of aquatic mammal exploitation also fall into the archaic period, although not from the proper Mesoamerican cultural area itself. Guía-Ramírez's analysis of archaic exploitation of sea mammals shows how hunter-gatherer groups from Aridamerica were able to feed especially on pinnipeds, still with rudimentary technologies.

The beginning of the Preclassic or Formative period (approx. 4,000 BP – 1,800 BP), is situated in a moment when the Olmec culture played a dominant role (Coe & Diehl 1980), and sites with its characteristics are distributed throughout the area (Clark & Pye 2000). Important sites in the center of Mesoamerica mark the formal and autochthonous development of the Preclassic period in central Mexico and also exhibit the dense settlements and

[1] The Archaic period roughly corresponds to the Superior Cenolithic (approx. 9,000 – 7,000 BP) and Protoneolithic period (approx. 7,000 – 4,500 BP) in Mexican Archaeology.

formal architecture of this period (Clark et al. 2000; Grove 2006). It is especially during the establishment and initial development of these last settlements, along with Monte Albán and Teotihuacan in the Mexican highlands, that the formation of states can be considered to have begun, near 2,500 BP (Smith & Masson 2000:8). The chapter of Corona, Carmona and Arroyo indicates a particular find of a cetacean rib that belongs to the cultural sphere of the Classic period in southwestern Mesoamerica. During the Classic period (approx. 2,200 BP – 1,100 BP), most of Mesoamerica was likely occupied by high civilizations, several of these dominated and many more strongly influenced by the large and powerful city of Teotihuacan (Manzanilla 2001).

The classic Maya cities of the central lowlands of southeastern Mesoamerica appear to have been clearly tied to each other, but were also marked by a strong competitiveness (Martin & Grube 2008) and heterogeneity in cultural patterns and adaptations to local environments (Benavides 2001a; Sharer 1994). The chapter from Götz, Valadez and Rodríguez refers to this moment of cultural development as well as to its chronological evolution, in which the Maya fisherman exploited especially the shallow waters of the Gulf of Mexico and the Caribbean Sea. Near the ninth century of our era, many central and southern lowland Maya cities seem to have lost political and economic power, some even populations, in a much debated process known as the 'Maya collapse' (Aimers 2007; Demarest et al. [eds.] 2004; Marcus 2001), for which diverse causes are discussed. The Maya area itself, despite social changes, never became depopulated, and the northern lowlands witnessed cultural growth, especially during the 10th and 11th centuries, corresponding to the Terminal Classic period (approx. 1,100 BP - 950 BP) (Benavides 2001b; Cobos 2004). Dense settlement patterns in the Maya zone of the Pu'uc and the strong development of cities like that Chichén Itzá testify to this cultural florescence (Cobos 2004). A similar florescence was seen in central Mexico with the growth of multiple competitive cities, such as Xochicalco or Cacaxtla, during the Epiclassic period, roughly corresponding to the Terminal Classic period of the Maya area (Foncerrada 1993; Hirth & Cyphers 1988). The decline of Teotihuacan as a dominant site of Mesoamerica during the Classic had left a vacuum of power, which motivated the formation of cultural mosaics that marked the transition to Mesoamerica's last Pre-Hispanic period, the Postclassic. Finally, with the conquest of Tenochtitlan in 1521, the Spanish dominion in Mesoamerica initiated the gradual but constant disappearance of the native Pre-Columbian cultures throughout the following one hundred years (Gunsenheimer 2000), causing thus the beginning of a new era.

On the other hand, the Antillean geographic area was occupied by waves of human migrations that arrived at different times since about 6,000-8,000 BP. In Cuba, characterization of these aboriginal communities has been treated by several authors, such as Rouse (1964, 1986), Pichardo (1956), Tabio & Guarch (1966), Tabío (1984) and Guarch (1990), being the latter two the most accepted by the community of Cuban archaeologists (Lourdes Pérez Iglesias, personal communication). As Wilson (2007:21) has recently summarized, the scheme developed by Cuban archaeologists is based on the economic systems of these ancient peoples and it consists of two stages of economic development: the economy of appropriation and the economy of food production.

The economy of appropriation stage recognizes three phases: hunting, fishing and collecting, and the protoagricultural phase, which chronologically speaking, covers between 10,000 and 700 BP (Lourdes Pérez Iglesias, personal communication). These phases corresponds to what in the classification of Chanlatte (1995) is called the Archaic period; within this, "Chanlatte defines a preceramic period in which there was little or no interaction with pottery-making groups outside the islands and a ceramic period of development that took place with some interaction between Archaic and people who made ceramics and practiced horticulture" (Wilson 2007:21).

The stage of economy of food production, on the other hand, corresponds basically to the agricultural phase, or to the *Período Agroalfarero* (farmer-potter period, in the classification of Chanlatte (1995), who "describes two migrations – Agroalfarero I and II – representing separate movements. From the interaction of these groups and Archaic people emerges the Antillean Formative, which includes all subsequent prehistoric people. Agroalfarero III includes people between about 460 and 1,025 AD., and Agroalfarero IV refers to the late Taíno fluorescence seen by Spanish chroniclers" (Wilson 2007:22).

Going south, we reach the particular case of what nowadays is Colombia. The main general chronological chart for the prehispanic period has been Reichel-Dolmatoff´s (1986) scheme, made of five major periods: Paleoindian Period, the Archaic Period, the Formative Period, the Regional Developments Period, the Chiefdom´s Period, and the Incipient States Period. In this scheme, the transition between the first and the second periods is basically marked by the introduction of horticulturalist practices which in turn will pave the way for the Formative period which, in turn, is subdivided into lower, middle and upper phases. This subdivision marks the transit of relatively small scale communities based in the production of yucca to maize based societies, with villages, ceramics, gold metallurgy and more dense settlements. The last periods of this scheme represent more complex organizations some of them with major public works associated, such as the Taironas and Muiscas.

Reichel-Dolmatoff´s scheme, although it is still used as the major frame in general books and museographic scripts, is being questioned in several regards, as a result of advances in archaeological research intended to document the cultural diversity observed in the same periods of time across different regions of the country. Thus, as a result, a more compact scheme of just three major periods has been now considered more appropriated. In this "simpler" scheme, the first period is the Hunter-Gatherer and Horticulturalists period (20,000-16,000 to about 5,000 BC), a long interval of time in which according to recent discoveries, horticulturalist practices were an integral part of the subsistence strategies of early groups which exhibit a great variety of lithic, shell and bone artifacts matching the diversity of the Colombian territory (Jaramillo 2012; Gnecco & Aceituno 2004). The next period is the Formative Period (5,000 BC – 500 AD), in which overall evidence indicates that societies are now organized upon agriculturalist systems, particularly of maize, moving to more and more dense and stratified type of societies, including rather complex villages (for a discussion of the major changes that characterize this period, see Langebaek & Dever 2000). The last period before the Spanish arrival is the so called Regional Developments Period (500 – 1,500 AD), when in many regions evidence of complex chiefdoms are now in place and gold metallurgy and some major public works are also present (Drennan 2008; Langebaek 1995; Jaramillo 2008). Finally, it is important to point out that the scarce archaeological evidences for the use of aquatic mammals in Colombia come all from sites of the Formative Period.

Finally, regarding the South of the American continent, it should be said that the region followed a trend towards insularization after the end of the Pleistocene, when glacial melting and changes in sea level modeled strongly the Patagonian landscapes. Since that time, Patagonia has undergone a number of important geological changes, basically implying an accentuation of the characteristics derived from its peninsular shape. On the one hand, the rise of sea level resulted in the submergence of approximately one million square kilometres (Iriondo 2000). On the other, Pleistocene plate tectonics and glacial activity ended up in the fragmentation of the resulting space, thereby generating numerous islands southwest and south of the continent (Rabassa et al. 2000).

By dispersing into South America, humans reached the southernmost latitudes much before recent times. The first human occupations in Patagonia occurred at the end of the Pleistocene and, hence, human arrival to this area dealt with changing ecological conditions like those mentioned in the previous paragraphs, which affected both human populations and their prey and competitors on the mainland as well as in the newly formed islands (see Legoupil 1993-1994; Borrero 2002; Miotti & Salemme 1999, 2003). Climatic fluctuations implied in the Pleistocene-Holocene transition had an uneven impact upon Fuego-Patagonian animal populations. Some taxa, such as a horse (*Hippidion saldiasi*), a ground sloth (*Mylodon darwinii*), a panther (*Pantheraonca mesembrina*), became extinct, while others, such as the guanaco (*Lama guanicoe*), two South American foxes (*Pseudalopex culpaeus* and *P. griseus*), and the puma (*Puma concolor*), outlived these changes (Miotti & Salemme 1999, Borrero 2001). Concerning the human-animal relationship, the first human populations living in what is now Patagonia and Tierra del Fuego used a mostly modern fauna, camelids and carnivores in particular. It is yet to be determined whether this prevalence of modern taxa relates to a higher abundance of these prey animals, or whether it is the result of a specific human predatory strategy on this kind of fauna (see Miotti & Salemme 1999, Borrero 2001). The archaeological record suggests the use of generalized strategies oriented to terrestrial animals at the end of the Pleistocene, in which guanacos and, to a lesser extent, horses would have been the main resources (Borrero & Franco 1997, Miotti & Salemme 1999). Marine mammals are not important in the archaeofaunal records until 7.000-6.000 BP in south of Tierra del Fuego (Schiavini 1993, Orquera & Piana 1999). Pinnipeds, in particular have been relevant to patagonian human populations throughout the Holocene for different reasons, among them, the amounts of fat they offer, which is a critical substance at these latitudes; the energetic subsidy from the marine to the terrestrial environment they represent (Schiavini 1993), and their predictability in time and abundance in certain areas of the coastal landscapes.

According to the ethnographic record, there were marine hunter-gatherers west and south of the Patagonian archipelago, and terrestrial hunter-gatherers in the continental sectors. The former actively occupied the channels around the island by means of canoes. The latter, instead, were considered pedestrian hunters. According to this view, both groups shared nomadism and the simplicity of their technologies. This ethnographically derived classification was dominant for a long time since its introduction, and was often used in a dichotomic fashion. More recently, several studies have shown that in pre-ethnographic times human strategies in what is now Tierra del Fuego and Patagonia were more varied than previously thought. Research in different lines of evidence, such as stable isotopes, show a more diverse scenario in the exploitation of marine resources than suggested by the ethnographic record all through southern Patagonia (see Muñoz, 2011, for a review). Regarding pinnipeds, the main marine mammals in prehistoric human economies, such variation involves different strategies, ranging from the sporadic, occasional exploitation of these mammals to their intensive and specialized use by hunter-gatherers. In turn, these diverse strategies should not be regarded as static, but variable through time.

In the following paragraphs, a brief description of the contribution of the different chapters of this book is presented, clarifying that for us, one of the points that strongly connect all of the chapters is the importance that zooarchaeological data can have to biological conservation discussions. Figure 1 shows the areas covered by the book, from the extreme north to the southernmost corner of the Neotropical region.

Figure 1. Map of the American continent, showing the areas treated by the chapters of this book. Drawing by Christopher M. Götz.

Guía-Ramírez considers the exploitation of the early, e.g., archaic hunters of Aridamerica's west coast at Baja California. Although aquatic mammal hunt and exploitation does not seem to have been the main subsistence activity of the early hunter-gatherer groups of that area, archaeological remains of those animales can be frequently found in the archaeological context. Guía-Ramírez analyzes the taxonomic abundance of sea mammals at archaeological sites, as well as the artifactural technology that apparently was used to pursue those taxa, and reaches interesting conclusions about the temporal variations of marine mammal hunting in Baja California.

Götz, Valadez and Rodríguez deal with the archaeological and ethnohistorical evidence of aquatic mammals from the Maya culture of Mesoamerica's southeastern limits. It is known that the Maya exploited and used aquatic mammals for a variety of purposes, among which the use for food, but also as prime material for artifacts and sculptures was most prominent. Numerous archaeological sites in the Maya area have revieled remains of manatees, but also West Indian seals and dolphins. In spite of the relative frequency of aquatic mammal bone and teeth finds at Maya archaeological sites, it can also be deduced that the prehispanic Indian population of this region did not severely impact or even extinct those marine animals, since populations continued to be numerous until well into colonial times.

Corona, Carmona and Arroyo describe a particular find made deep inland in southern Mesoamerica, but probably related to the west coast of Oaxaca. At the site of Monte Albán, an important Zapotec center adyacent to the Mexican Highlands, an archaeological whale rib was found, belonging to a set of structures of the site´s center. The rib, well described by the authors, could have been a music instrument and was probably played in ceremonies related to sacrifices and mortuary offerings. Although it cannot be determined exactly from where the rib came from, a proveniance from the Pacific coast of southern Mesoamerica is probable.

Ramos Roca and Pérez Iglesias´s chapter reviews and discusses the available information on the use of marine mammals drawn from archaeological reports from the Caribbean and northern South America and in particular from Colombia, Cuba and the Antilles. The Antilles particularly woke the interest of specialists, especially in relation to the effects that over-exploitation by humans has had on these species, some of which are now extinct, while others face extinction or are potentially endangered. The chapter discusses the usefulness of this information for reconstructing the uses marine mammals in the region. In addition to confirming the scarcity of data on this group of vertebrates the chapter provides evidence of how in some cases the indigenous population considered their capture to be of particular importance, because of the abundant edible biomass they provided and the raw materials extracted from them that were used to manufacture body ornaments and funerary and ceremonial objects.

Carvajal's chapter summarizes the use of aquatic mammals by human populations in northern South America (Colombia and Panama) from ethnographic and ethnohistoric accounts from the 16th to 19th centuries. It discusses the availability and access of aquatic mammals for people based on ethnohistoric sources and also uses ethnographic information to illustrate how people hunted and consumed meat and fat from these aquatic mammals. In addition, the written records are employed to illustrate how aquatic mammals were used to manufacture tools and adornments.

Borella offers a complete review of the methodological status of sea lion studies in continental Patagonia (Argentina). Her analysis highlights the need to develop and incorporate an array of different analytical methods – osteometric studies, bone fusion chronology, identification of bone diagnostic traits – in order to get a more accurate picture of the role pinnipeds played in the subsistence strategies of Patagonian past human populations.

Zangrando discusses long-term variation in the taxonomical composition of pinniped zooarchaeological assemblages from insular Patagonia, namely the Beagle Channel region of Tierra del Fuego Island. He discusses new zooarchaeological data that helps to understand harvest strategies of sea lions through time in the Beagle Channel region, Southern Patagonia. Results show that a more variable scenario than previously thought is arguable, leading to a new round of more precise questions about the relationship between humans and pinnipeds in the extremes of the continent.

Finally, Tunez and Pimper offer a description of the rich marine mammals diversity in southern South America´s marine ecosystems and analyze the recorded presence and conservation status of six of the most characteristic species found in southern Neotropical coasts and waters.

ACKNOWLEDGMENTS

As editors of this volume, we would like to first thank the editorial Nova Science for the patience and support during the time of development of the manuscript. Gathering a manuscript about this theme required a considerable amount of time and effort, since there are, throughout Latin America, relatively few studies and investigators working with zooarchaeological finds in the first place, and marine/aquatic mammal remains in the second. Nevertheless, an important group of researchers agreed to publish with and we'd like to extend our warmest thanks to those authors, who agreed to send us their important contributions to the field; only through the continuous research effort of those zooarchaeologists and biologist who decidedly work with aquatic and especially sea mammal remains, could we hope to establish this publication. Of course, one of the main pillars of the academic value of this edition were the anonymous reviewers, who with much detail reviewed and commented the various chapters, and much of the updated information presented in this volume is due to their expert reviews and critical analysis of the contributions. Also, the different organisms which provided additional information of the chapters are acknowledged here, among those is the CONABIO, which, through Laura Cárdenas, permitted Götz, Valadez and Rodríguez to use images of their important digital database available online. We also want to thanks Cuban zooarchaeologist Lourdes Pérez, for the information related to cultural development on Cuba and the Antilles.

REFERENCES

Adams, R. E. W. (1991). *Prehistoric Mesoamerica.* Norman, USA: University of Oklahoma.

Aimers, J. J. (2007). What Maya Collapse? Terminal Classic Variation in the Maya Lowlands. *Journal of Archaeological Research 15(4)*, 119-377.

Benavides, A. (2001a). El norte de la zona maya en el clásico. In Manzanilla Naim L. & L. López Luján (Eds.), *Historia Antigua de México, El Horizonte Clásico vol. II* (pp. 119-160). Mexico City, Mexico: Instituto Nacional de Antropología e Historia, Universidad Nacional Autónoma de México, Editorial Porrúa.

Benavides, A. (2001b). El sur y el centro de la zona maya en el clásico. In Manzanilla Naim L. & L. López Luján (Eds.), *Historia Antigua de México, El Horizonte Clásico vol. II* (pp. 79-118). Mexico City, Mexico: Instituto Nacional de Antropología e Historia, Universidad Nacional Autónoma de México, Editorial Porrúa.

Borrero, L. A. (2001). *El Poblamiento de la Patagonia.* Buenos Aires, Argentina: Emecé Editores.

Borrero, L. A. (2002). The Archaeology of Patagonia. In Briones, C. & J. L. Lanata, (Eds). *Archaeological and anthropological perspectives on the native peoples of Pampa, Patagonia and Tierra del Fuego,* (pp. 47-55). London, UK: Bergin & Gargey.

Borrero, L. A. & Franco, N.V. (1997). Early Patagonian Hunter-Gatherers: Subsistence and Technology. *Journal of Anthropological Research 53*, 219-239.

Chanlatte, L. (1995). Presencia Huecoide en Hacienda Grande, Loiza, Puerto Rico. In *Proceedings of the Fifteenth International Congress for the Study of Pre-Columbian Cultures of the Lesser Antilles* (pp. 501– 515). San Juan, Puerto Rico: Centro de Estudios Avanzados de Puerto Rico y el Caribe.

Clark, J. E. & Pye, M. E. (2000). *Olmec Art and Archaeology in Mesoamerica.* New Haven, USA: Yale University Press.

Clark, J. E., Hansen, R. & Pérez Suárez, T. (2000). La zona Maya en el preclásico. In Manzanilla Naim L. & L. López Luján (Eds.). *Historia Antigua de México El México Antiguo, sus Áreas Culturales, los Orígenes y el Horizonte Preclásico, Vol 1* (pp. 437–510). Mexico City, Mexico: Instituto Nacional de Antropología e Historia, Universidad Nacional Autónoma de México, Editorial Porrúa.

Cobos Palma, R. (2004). Chichén Itzá: Settlement and Hegemony during the Terminal Classic Period. In Demarest, A. A., P. M. Rice, & D. S. Rice (Eds.). *The Terminal Classic in the Maya Lowlands: Collapse, Transition, and Transformation* (pp. 517–544). Boulder, USA: University Press of Colorado.

Coe, M. D. & Diehl R. A. (1980a). *In the Land of the Olmec: The Archaeology of San Lorenzo Tenochtitlan: Vol. 1.* Austin, USA: University of Texas Press.

Demarest, A. A., Rice, P. M. & Rice, D. S. (Eds) (2004). *The Terminal Classic in the Maya Lowlands: Collapse, Transition, and Transformation.* Boulder, USA: University Press of Colorado.

Drennan, R. (2008). Chiefdoms of southwestern Colombia. In Silverman, H. & W. Isbell (Eds.). *Handbook of South American Archaeology* (pp. 381-403). New York, USA: Springer.

Foncerrada de Molina, M. (1993). *Cacaxtla, La Iconografía de los Olmeca-Xicalanca.* Mexico City, Mexico: Universidad National Autónoma de México, Instituto de Investigaciones Estéticas.

García Bárcena, J. (2001). Cenolítico superior y protoneolítico (7000 a.C.– 2500 a.C.). *Arqueología Mexicana 9 (52)*, 52–57.

Gnecco, C. & Aceituno, J. (2004). Poblamiento temprano y espacios antropogénicos en el norte de Suramérica. *Complutum 1*, 151-164.

Grove, D. (2006). Chalcatzingo y la "Cultura Tlatilco" en el preclásico de Morelos. In López Luján, L., D. Carrazco & L. Cue (Eds.). *Arqueología e Historia del Centro de México: Homenaje a Eduardo Matos Moctezuma* (pp. 101–102). Mexico City, Mexico: Instituto Nacional de Antropología e Historia, Universidad Nacional Autónoma de México.

Guarch, J.M. (1990). *Estructura para las Comunidades Aborígenes de Cuba.* La Habana, Cuba: Ediciones Holguín.

Gunsenheimer, A. (2000). Entre adaptación y rebelión: La sociedad Maya durante el período colonial (1546–1811). In Grube, N. (Ed.). *Los Mayas: Una Civilización Milenaria* (pp. 384–395). Hagen, Germany: Könemann.

Haberland, W. (1991). *Amerikanische Archäologie.* Darmstadt, Germany: Wissenschaftliche Buchgesellschaft.

Hirth, K. G. & Cyphers, A. (1988). *Tiempo y Asentamiento en Xochicalco.* Mexico City, Mexico: Universidad Nacional Autónoma de México, Instituto de Investigaciones Antropológicas.

Iriondo, M. (2000). Patagonian dust in Antarctica. *Quaternary International 68/71*, 83-86.

Jaramillo, L. G. (2008). *Sociedades Prehispánicas en el Territorio "Quimbaya": Unidades Domésticas, Áreas de Actividad y el Complejo Tesorito.* Bogotá, Colombia: Fundación de Investigaciones Arqueológicas Nacionales del Banco de la República.

Jaramillo, L. G. (2012). Entre "grandes esquemas" y "escalas pequeñas": reflexiones metodológicas sobre los estudios alimentarios y de vida doméstica en el altiplano cundiboyacense (Colombia). In Babot, M. P., M. Marschoff & F. Pazzarelli,(Eds.). *Las Manos en la Masa: Arqueologías, Antropologías e Historias de la Alimentación en Suramérica* (pp. 93-118). Córdoba, Argentina: Universidad Nacional de Córdoba. Facultad de Filosofía y Humanidades; Museo de Antropología UNC - Instituto Superior de Estudios Sociales UNT, Argentina.

Kirchhoff, P. (1967). *Mesoamérica: Sus límites Geográficos, Composición Étnica, y Caracteres Culturales.* Mexico City, Mexico: Universidad Nacional Autónoma de México, Escuela Nacional de Antropología e Historia.

Langebaek, C. (1995). *Regional Archaeology in the Muisca Territory: a Study of the Fúquene and Susa Valleys /Arqueología Regional en el Territorio Muisca: un Estudio de los Valles de Fúquene y Susa. University of Pittsburgh Memoirs in Latin American Archaeology No. 9.* Pittsburgh – Bogotá, USA & Colombia: University of Pittsburgh, Department of Anthropology - Universidad de los Andes, Departamento de Antropología.

Langebaek, C., & Dever, A. (2000). *Arqueología en el Bajo Magdalena: Un Estudio de los Primeros Agricultores del Caribe Colombiano. Informes Arqueológicos del Instituto Colombiano de Antropología e Historia Nº 1.* Bogotá, Colombia: Instituto Colombiano de Antropología e Historia y Universidad de los Andes.

Legoupil, D. (1993-94). El archipiélago del Cabo de Hornos y la costa sur de la isla Navarino: poblamiento y modelos economicos. *Anales del Instituto de la Patagonia, Serie Ciencias Humanas 22*, 101-121.

Lohse, J. C., Awe, J., Griffith, C. Rosenswig, R. M. & Valdez, F. Jr. (2006). Preceramic Occupations in Belize: Updating the Paleoindian and Archaic Record. *Latin American Antiquity 17 (2)*, 209–226.

Manzanilla, L. (2001). La zona del Altiplano central en el clásico. In Manzanilla Naim L. & L. López Luján (Eds.). *Historia Antigua de México El México Antiguo, El Horizonte Clásico Vol II* (pp. 203–239). Mexico City, Mexico: Instituto Nacional de Antropología e Historia, Universidad Nacional Autónoma de México, Editorial Porrúa.

Marcus, J. (2001). La zona Maya en el clásico terminal. In Manzanilla Naim L. & L. López Luján (Eds.). *Historia Antigua de México El México Antiguo, El Horizonte Clásico Vol II* (pp. 301–346). Mexico City, Mexico: Instituto Nacional de Antropología e Historia, Universidad Nacional Autónoma de México, Editorial Porrúa.

Martin, S. & Grube, N. (2002). *Crónicas de los Reyes y Reinas Mayas.* Mexico City, Mexico: Planeta.

Masson, M. & Peraza Lope, C. (2008). Animal Use at the Postclassic Maya Center of Mayapan. *Quaternary International 191*, 170–183.

Matos Moctezuma, E. (1994). Mesoamérica. In Manzanilla Naim L. & L. López Luján (Eds.). *Historia Antigua de México El México Antiguo, sus Áreas Culturales, los Orígenes y el Horizonte Preclásico Vol 1* (pp. 49–73). Mexico City, Mexico: Instituto Nacional de Antropología e Historia, Universidad Nacional Autónoma de México, Editorial Porrúa.

Mirambell S., L. (1994). Los primeros pobladores de actual territorio mexicano. In Manzanilla Naim L. & L. López Luján (Eds.). *Historia Antigua de México El México Antiguo, sus Áreas Culturales, los Orígenes y el Horizonte Preclásico Vol 1* (pp. 177–208). Mexico City, Mexico: Instituto Nacional de Antropología e Historia, Universidad Nacional Autónoma de México, Editorial Porrúa.

Miotti, L. & Salemme, M. (1999). Biodiversity, taxonomic richness and specialists-generalists during Late Pleistocene/Early Holocene times in

Pampa and Patagonia (Argentina, Southern South America). *Quaternary International 53/54*, 53-68.

Miotti, L. & Salemme, M. (2003). When Patagonia was colonized: people mobility at high latitudes during Pleistocene/Holocene transition. *Quaternary International 109-110*, 95-111.

Mirambell S., L. (2001). Arqueolítico y cenolítico inferior (30000–7000 a.C.). *Arqueología Mexicana* IX (52):46–51.

Morello, J. (1984). *Perfil ecológico de sudamérica. Características Estructurales de Sudamérica y su Relación con Espacios Relevantes del Planeta Vol 1*. Barcelona, España: Ediciones Cultura Hispánica, Instituto de Cooperacion Iberoamericana.

Muñoz, A. S. & Mondini, M. (eds.) 2008. Neotropical Zooarchaeology and Taphonomy. *Quaternary International 180.*

Nárez, J. (2000). Aridamérica y Oasisamérica. In Manzanilla Naim L. & L. López Luján (Eds.). *Historia Antigua de México El México antiguo, sus áreas culturales, los orígenes y el horizonte preclásico Vol 1* (pp. 121–158). Mexico City, Mexico: Instituto Nacional de Antropología e Historia, Universidad Nacional Autónoma de México, Editorial Porrúa.

Niederberger, C. (1996). Paisajes, economía de subsistencia y agrosistemas. In Lombardo de Ruíz, S. & E. Nalda (Eds.). *Temas Mesoamericanos* (pp. 11–50). Mexico City, Mexico: Instituto Nacional de Antropología e Historia & CONACULTA.

Orquera, L. A. & Piana, E. L. (1999). *Arqueologia de la Región del Canal Beagle (Tierra del Fuego, Republica Argentina).* Buenos Aires, Argentina, Publicaciones de la Sociedad Argentina de Antropología.

Pichardo, F. (1956). *Los Aborígenes de las Antillas.* Mexico City: Mexico, Fondo de Cultura Económica.

Prem, H. J. (1989). *Geschichte Altamerikas.* Munich, Germany: Oldenbourg.

Pohl, M. E. D., Piperno, D. R. Pope, K. O. & Jones, J. G. (2007). Microfossil Evidence for Pre-Columbian Maize Dispersals in the Neotropics from San Andrés, Tabasco, Mexico. *Proceedings of the National Academy of Sciences 104 (16)*, 6870–6875.

Rabassa, J., Coronato, A., Bujalesky, G., Salemme, M., Roig, C., Meglioli, A., Heusser, C., Gordillo, S., Roig, F, Borromei, A. & Quattrocchio, M. (2000). Quaternary of Tierra del Fuego, southernmost South America: an updated review. *Quaternary International 68/71*, 217:240.

Redford, K. H. & Eisenberg, J. F. (1992). *Mammals of the Neotropics, The Southern Cone Vol 2*. Chicago, USA, Chicago University Press.

Reichel-Dolmatoff, G. (1986). *Arqueología de Colombia: Un Texto Introductorio*. Bogotá, Colombia: Fundación Segunda Expedición Botánica - Departamento Administrativo de la Presidencia de la República. Litografía Arco.

Rouse, I. (1964). Prehistory of the West Indies. *Science 144 (3618)*, 369–375.

Rouse, I. (1986). *Migrations in Prehistory*. New Haven, USA: Yale University Press.

Schiavini, A. (1993). Los lobos marinos como recurso para cazadores-recolectores marinos: el caso de Tierra del Fuego. *Latin American Antiquity 4*, 346-366.

Sharer, R. J. (1994). *The Ancient Maya. 5th ed.* Stanford, CA., USA: Stanford University Press.

Smith, M. E. & Masson, M. A. (Eds.), (2000). *The Ancient Civilizations of Mesoamerica: A Reader.* Malden, USA, Blackwell.

Tabío, E. & Guarch, J. M. (1966). *Excavaciones en Arroyo del Palo, Mayari, Cuba*. Havana, Cuba. Departamento de Antropología, Academia de Ciencias de la República de Cuba.

Tabío, E. (1984). Nueva Periodización para el estudio de las comunidades aborígenes de Cuba. *Islas 78*, 35-51.

Valadez Azúa, R. (2003). *La Domesticación Animal.* Mexico City, Mexico: Universidad Nacional Autónoma de México, Instituto de Investigaciones Antropológicas.

Wilson, S. (2007). *The Archaeology of the Caribbean. Cambridge World Archaeology Series*. Cambridge, USA: Cambridge University Press.

In: Neotropical and Caribbean Aquatic Mammals ISBN: 978-1-63321-306-7
Editors: Muñoz, Götz and Ramos Roca © 2014 Nova Science Publishers, Inc.

Chapter 1

MARINE MAMMALS IN ARCHAEOLOGICAL CONTEXTS ON THE BAJA CALIFORNIA PENINSULA

Andrea Guía-Ramírez*
Centro Regional de Baja California del Instituto Nacional de Antropología e Historia, México

ABSTRACT

The present chapter describes the current situation of archaeological finds that evidence levels of early exploitation of marine mammals in Baja California, Mexico. Three regions are scrutinized in the text--northern, central and southern Baja California--where sites with Archaic (7,500 – 1,500 BP) and Late Prehistoric (1,500 - 410 BP) occupation showed varying degrees of sea mammal hunting and exploitation levels. The archaeological remains, mainly from sea lion *Zalophus californianus*, sea otter *Enhydra lutris*, and cetaceans, indicate that the animals were occasionally hunted, with technologies similar to those in ethnohistorical accounts, and that temporal variations occurred, limiting the numbers of hunted sea mammals towards the Late Prehistoric period, which ended with Spanish conquest.

* ghia.ramirez@gmail.com

INTRODUCTION

The Baja California peninsula is situated between the coordinates 32°32′03” and 22°51′15” north and between 117°07′38” and 109°54′14” west. It is a long, narrow arm of land with a length of approximately 1,260 km and a maximum width of 175 km. It is bounded on the east by the Colorado River, which empties into the warm waters of the Gulf of California, and on the west by the Pacific Ocean. To the north, it is connected with the continental mass through the states of Sonora in Mexico and California and Arizona in the U.S. Throughout its extension, one of the most diverse and contrasting environmental scenes in Mexico is observed. Its interior is a semi-desert region, dominated by low hills with valleys, mountain ranges, and arid deserts. The peninsula's physiographic differences manifest themselves in varied climates along its length and width. The northwest has a Mediterranean climate. In the mountains, snowfall is recorded for the winter. The central portion shows an extreme aridity that crosses obliquely from the Colorado delta to the Pacific coast. In the south, tropical and subtropical climates are present (Maya et al. 2011).

On the peninsula, about 2,500 archaeological sites have been identified. Some of them attest to a human presence on the peninsula that goes back a little more than 10,000 years, into the late Pleistocene (Des Lauriers 2006; Fujita 2006; Gutiérrez & Hyland 2002). The archaeological record reveals a predominance of native groups with a hunter-gatherer-fisher mode of life, with relatively simple technological systems and a social structure based on the universal concepts of age and sex (Laylander 2006).

The extent of coastal areas and the exploitation of marine resources allowed the formation of the conspicuous sites termed shell middens. These are defined as archaeological sites formed by the remains of mollusks (shells) and other refuse, which indicate subsistence based on marine resources through collecting of shellfish, fishing, and exploiting marine mammals. The study of shell middens offers a panoramic view of the means of subsistence of the peninsula's ancient inhabitants, the natural resources to which they had access, their networks of exchange, human-environment relations, and climatic variations that influenced the presence and distribution of faunal and floral resources, as well as the groups' ways of adapting to changing climatic conditions.

Today, between 45 and 49 species of marine mammals are known in Mexico, distributed in 12 different families (Medrano et al. 2007). Eleven of the 12 families represented in Mexico inhabit the waters of the Pacific Coast of Baja California, while eight families are recorded in the Gulf of California. The first of these zones is the richest area in Mexico and counts 35 species, while the second, the Gulf of California, records 32 of the species reported for Mexico (Torres et al. 1995). The great diversity and enormous richness of the resources available in the marine ecosystems is beyond doubt. Torres and collaborators (1995) consider that, among undomesticated mammals, marine mammals are the ones most used by man, with a long history in terms of their capture, for some species leading to their reduction, fragmentation of ranges, and extinction.

The role of marine mammals has been studied at archaeological sites in diverse parts of the world, as they were exploited in numerous coastal societies. In these studies, the discussion usually focuses on the contribution of this faunal group to the subsistence and cultural development of human societies linked to marine resources (Jones et al. 2002; Lyman 1989, 1992). Although marine mammals typically provided important resources in various forms such meat, hides, fat, and raw materials, they did not always constitute the principal resource in the diet of the human groups that exploited them, particularly at sites in the temperate zones (Colten & Arnold 1998). However, the record includes a few societies that based their economy on the exploitation of this resource, making it their principal food source (Renker & Gunther 1990; Sheehan 1985).

Numerous studies conducted by different investigators at sites near coastal shorelines have provided important data for the study of these mammals in cultural contexts. Several sites on the Pacific Coast, radiocarbon-dated to the late Holocene, show evidence of temporal variation both in the size and in the faunal composition of the pinniped community, which is explained as a response to the disturbance brought about by the hunting of these animals (Lyman 1995; Porcasi et al. 2000; Walker et al. 2000). This model is shown to be consistent at several archaeological sites distributed along the continental coastline, suggesting an increase in the economic importance of pinniped meat during the late Middle period (1,450-500 BP) (Jones et al. 2002)

After the excavation of numerous shell middens at coastal sites in California and Oregon in the U.S., it was determined that the species *Callorhinus ursinus* appears as one of the principal components at these sites. Based on the relative abundance of the remains of this mammal, which has been described at early sites and extended through a long period, it is suggested that it was a principal component of the continental coast's pinniped communities, in locations where it is rare to encounter it today. This is explained as an alternative for the ecological niche adopted by populations of

North Pacific fur seals, following pressure exerted by groups during the late Holocene. Some well-documented cases show that prehistoric man was able to wipe out local populations of marine mammals in only a few hundred years, once he arrived to colonize a new area (Gifford-Gonzalez 2004; Hildebrandt & Jones 1992). On the other hand, some researchers have tried to show that environmental changes were the determining factors in the variation in the composition of marine mammals within archaeological contexts (Colten 1995; Colten & Arnold 1995, 1998).

At other California sites, excavations dated to the Archaic (7,000-1,500 BP) and Late Prehistoric (1,500-418 B.P.) periods reflect changes in settlement and methods of subsistence through time. Variation in the use of coastal resources is shown, motivated by an adaptation to environmental changes from a regime of greater humidity to a drier and more hostile one, resulting in a search for alternative resources (Wake 1999).

Marine Mammals in Archaeofaunal Collections

Along the Baja California peninsula, numerous archaeological sites have been located, both on the coasts and in the interior. The coastal sites seem particularly conspicuous (Laylander 2006) and more densely populated (Oviedo-García and Guía-Ramírez 2009). At these sites, subsistence was based, to a greater or lesser extent, on the collection of mollusks (Drakic et al. 2005, 2006; Fujita & Poyatos 1998; Oviedo-García 2008; Porcayo 2008; Ritter 1998a), on fishing (Guía-Ramírez 2008a), and on the exploitation of marine mammals. At some sites, a mixture in the use of marine and terrestrial resources is attested (Fujita 2006; Oviedo-García & Guía-Ramírez 2009; Ritter 2006a), while others show a clear preference for marine exploitation (Des Lauriers 2006; Fujita 2006; Ritter 1998a, 2006a).

While remains of marine mammals are consistently present at the archaeological sites linked to the Baja California peninsula's coasts, they do not usually dominate their archaeological contexts (Guía-Ramírez 2005, 2010). The exceptions are certain sites where a very effective, perhaps even specialized, exploitation of this resource has been detected (Des Lauriers 2006; Porcasi & Fujita 2000).

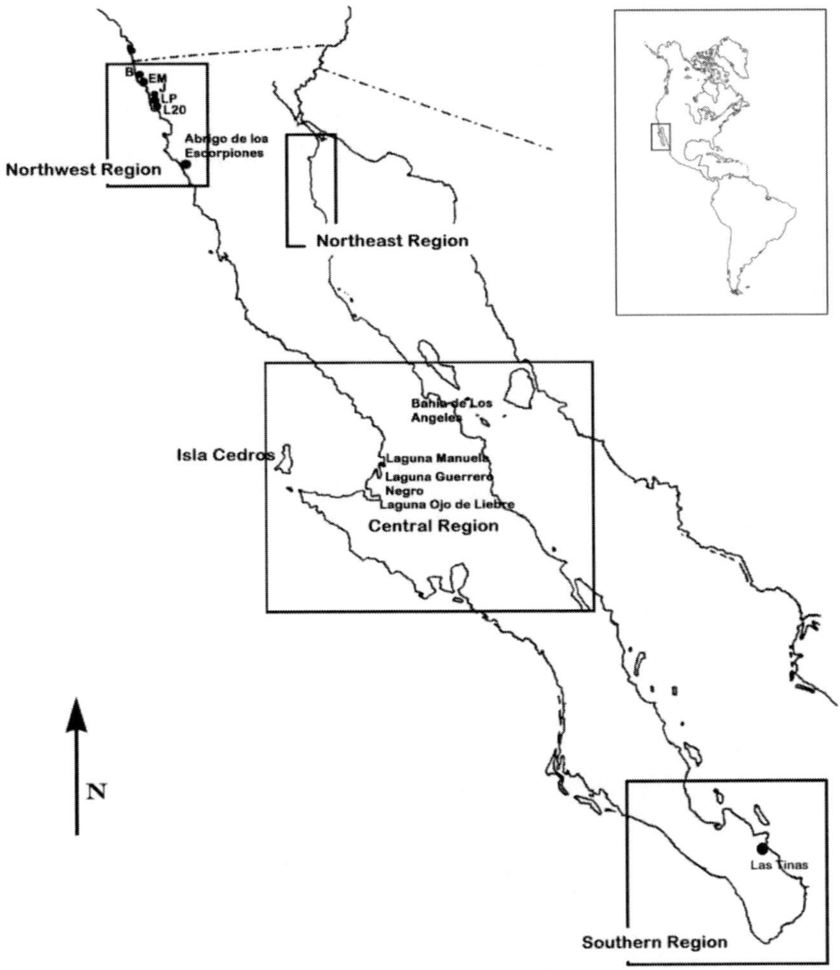

Figure 1. Map of the study area, indicating regions mentioned in the text. Drawing by the author.

Throughout the peninsula, there are radical variations in the type and availability of marine resources (Figure 1). Overall, the northwest portion is characterized by cold-water species, where archaeological sites are characterized as large, dense areas formed in particular by mussel (*Mytilus californianus*) and black abalone (*Haliotis cracherodii*), as well as the remains of fish, among which California sheephead (*Semicossyphus pulcher*), rockfish (*Sebastes*), and surfperches (Embiotocidae) are conspicuous. Some birds and terrestrial mammals are also recorded (Guía-Ramírez 2008a). Remains of marine mammals are scarce, constituting a small percentage of the faunal assemblage. Up to now, the presence of California sea lions (*Zalophus californianus*), sea otters (*Enhydra lutris*), the family Delphinidae, the order Cetacea, the family Balaenopteridae, and diverse elements classified under the heading of pinnipeds have been successfully identified (Guía-Ramírez 2005; Sánchez 2009). In general, California sea lions and sea otters are the species that occur most frequently in this region (Figure 2). Bone elements record evidence of thermal alteration from different degrees of exposure to fire and cut marks related to the task of disarticulation and defleshing (Figure 3). To a lesser degree, there is incipient evidence of raw material working. Remains of cetaceans are rare, and most are found to have been modified, worked as tools or as luxury items (Guía-Ramírez 2005).

The sites in the northeastern region are large concentrations of warm-water mollusks, in which various species of oysters of the genus *Ostrea*, bittersweet (*Glycymeris gigantea*), and fish remains of the Scianidae family are conspicuous (Celis 2011; Guía-Ramírez 2009). Marine mammals are scarce, limited to some fragments of bone, and at some sites their presence is not recorded (Guía-Ramírez 2007). However, this perception may change as archaeological studies develop in the region.

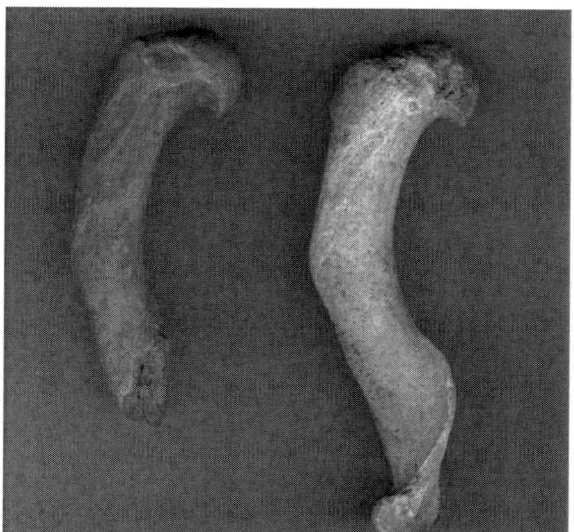

Figure 2. Humeri of *Enhydra lutris* from L 20 site. Photo by the author.

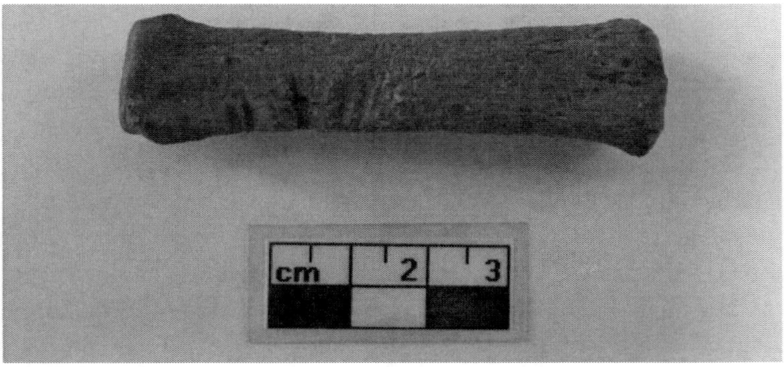

Figure 3. Phalanx from *Zalophus californianus* with butchering marks. Photo by the author.

In the central part of the Pacific coast, at the sites of the Three Sisters Lagoons (Laguna Manuela, Laguna Guerrero Negro, and Laguna Ojo de Liebre), some sites are formed primarily by speckled scallop (*Argopecten circularis*) and clams of the genus *Chione*, as well as the remains of diverse fish that were abundant and may have been the principal source of food. Also present are remains of birds and terrestrial mammals such as hares and rabbits. A meager presence of marine mammals is recorded, represented by harbor seal (*Phoca vitulina*), sea otter, and a juvenile whale. At these sites, the consumption and use of whales was apparently minimal and perhaps represented only the opportunistic exploitation of beached animals (Ritter 2006a).

In the central region on the Gulf of California, there are shell middens dominated by clams of the genera *Chione, Protothaca*, and *Cardita* (Bendímez et al. 1993; Ritter 1998a). There are also remains of rays, sharks, and fish from sandy substrates and reefs, along with the presence of remains from marine turtles and marine mammals (Ritter 1998c). Ritter (2006c) mentions that at Bahía de Los Ángeles, the exploitation of marine resources, principally pelagic and reef fisheries, was important. Based on the current presence of congregations of sea lions, he considered that their hunting ought to have been a determining factor for the sites in this area. While acknowledging the presence of marine mammals, no description of the species, genera, or families present has been made, although it is possible that the remains represent members of the order Cetacea and some pinnipeds.

The southern part of the peninsula down to the Cape region is characterized by sites formed by the remains of pearl oyster (*Pinctada mazatlanica*), rock oyster (*Ostrea*), and conch (*Strombus*), although this composition varies in adjacent regions and these species may be replaced by others (Fujita 2006). At some sites, a preference prevailed for predominantly maritime exploitation, while at other sites an exploitation of marine resources complemented with terrestrial resources by means of hunting and collection is observed (Fujita 2006; Fujita & Poyatos 1998). On some of the islands, in general terms and with relatively scarce frequencies, the remains of marine mammals indicate the presence of the California sea lion, the family Delphinidae, the order Cetacea, and remains classified as pinnipeds (Fujita & Guía-Ramírez 2011). On the other hand, some mainland sites report an elevated number of mammal bones, including ones recorded as pinniped, mostly consistent with identification as sea lion, although the presence of Guadalupe fur seal (*Arctocephalus townsendi*) is not ruled out. Also noted are an elevated frequency of the family Delphinidae, the rare presence of the order Cetacea, and an elevated number of unidentified mammal bones, which by their characteristics may correspond to marine mammals, specifically dolphins (Porcasi & Fujita 2000).

The exploitation of marine mammals does not appear to have been the key element of subsistence on the coasts of the Baja California peninsula, but it is represented in practically all coastal archaeological sites in this region. Few sites show an effective and perhaps specialized hunting of these mammals. At one site, Las Tinas 3, located on the Gulf of California in the Cape region south of La Paz, an intensive exploitation of small cetaceans was detected. The resource was processed and consumed at this site, as indicated by the presence of processing locations where many burned remains of dolphins were found (Porcasi & Fujita 2000). Another area of intensive exploitation of marine mammals was Isla Cedros, located on the Pacific. According to Des Lauriers (2006), fish appear to dominate the faunal assemblage at some sites on Isla Cedros, based on numbers of specimens. However, hunting or simply processing the carcasses of marine mammals may have contributed as much as or even more than fishing, especially at several sites where there is clear evidence of exploitation of pinnipeds and cetaceans (whales and dolphins), whose remains are found burned and sometimes used as raw materials. Indeed, the author considers that the presence of rookeries or zones for the congregation of marine mammals could have been a strong motive for the formation of some nearby settlements.

According to their abundance, the three groups of sea mammals most represented in archaeological sites northwest of Baja California are the sea lion *Zalophus californianus*, the sea otter *Enhydra lutris*, and cetaceans. However in the region, the intensive hunting of these groups does not seem to have been a predominant activity. The California sea lion is a recurring species in archaeological sites (Figure 4). It is reported along both coasts and across a north-south axis, a situation that is not surprising, given that it has a wide distribution and that a number of individuals may be taken at a low energy cost, in contrast to cetaceans, whose hunting requires a much greater undertaking. Based on the existence of congregations of multiple sea lions along the rocky coasts, it is not strange that they were among the prey most prized by the peninsula's native groups, although not heavily hunted. They provide large quantities of meat and fat, in addition to being easy to access on land, mostly in the case of females and juveniles whose hunting is relatively simple (Colten 1995; Glassow 1997; Jones & Hildebrandt 1995).

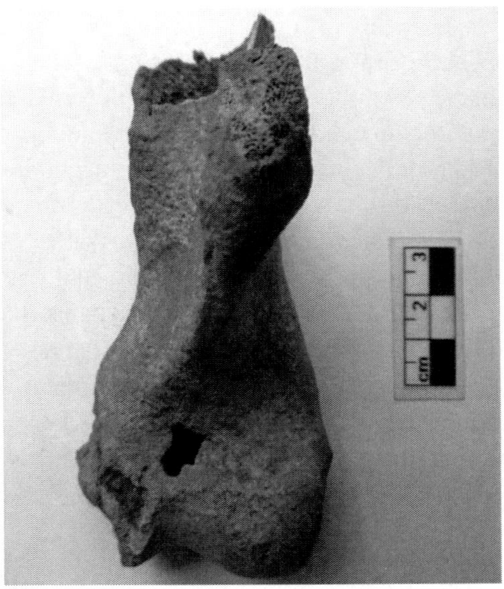

Figure 4. Humerus of *Zalophus californianus* from Bajamar site B. Photo by the author.

On the other hand, in accordance with its distribution, the sea otter appears to be the second most abundantly represented marine mammal in the archaeological sites of northwestern Baja California. Historically, this species inhabited the northern coast of the Pacific Ocean, from northern Hokkaido Island in Japan through the Commander and Aleutian Islands to Punta Morro in Baja California Sur (Zepeda-Gutiérrez 1986). Today, only a few colonies in California are known. Unlike sea lions, sea otters spend more time in the sea than on land, as a large part of their activity is focused on foraging in kelp beds (Kenyon 1969). Although the methods of hunting this species are not well documented in the native groups of Baja California, because of the behavior of the otter it is proposed a large effort was required to be able to hunt this species, including tools like harpoons, spears, and watercraft that led to where the animals were active. Nevertheless, it is also possible that they were captured in the short time they were on shore. In some sites in California, it is known that sea mammals were hunted in the ocean using harpoons, captured on the shore, and, in the case of the otter, using their brood as decoy. (McCawley 1996; Ogden 1941). This last hunting tactic is referenced in ethnographic data for groups of the peninsula (Lazcano 2000:77).

Cetaceans are also consistently represented in archaeological sites, although minimal in number. Except at certain sites, the exploitation of this resource would have been opportunistic (Des Lauriers 2006; Guía-Ramírez 2010a; Ritter 2006b), benefitting from the beaching of small and occasionally large cetaceans, which is common on the Baja California coasts. The use of beached animals was documented by the first explorers.) Pedro Porter, who explored the coast of Baja California in S. XVII, account that in Cabo San Lucas "a whale stranded on the coast, and in five days the Indians cut it to pieces with their hatchets, which are of stone" (González-Vázquez 2000). On the other hand the scarce archaeological evidence could support the lack of focused hunting of whales or dolphins, and might suggest more exploitation of individuals stranded on the beaches. Perhaps, as happens today, beaching occurred by chance, but these events were generally recurrent and numerous and they were exploited by scavenging, through which meat, fat, bone, and even bristles for the manufacture of tools and instruments were obtained with little effort.

Hunting Technology

The technology for hunting was deceptively simple but varied on the peninsula. At the majority of sites, there are no artifacts or objects exclusively for the hunting of marine mammals, such as the harpoon or lance (Oviedo-García, personal communication 2013). Nonetheless, at some sites it appears that the inhabitants were well equipped for undertaking that activity (Des Lauriers 2005, 2006). Still other groups achieved a highly specialized economy that reduced the technology and the energy costs, and the community organization and collective effort alone were sufficient for effective hunting of this resource (Porcasi & Fujita 2000).

Figure 5. Projectile points from northwest region of Baja California. Photo by Fernando Oviedo.

In Baja California, projectiles appeared to be sufficient artifacts for the acquisition of marine mammals, especially pinnipeds (Figure 5). According to numerous investigations carried out on the peninsula, projectile points are common and consistently present at archaeological sites. Although they were versatile tools used in the acquisition of diverse resources, the available archaeofaunal data (that is, the remains of marine mammals) by themselves may be sufficient evidence to suggest that these tools were also used for the procurement of this resource (Des Lauriers 2006; Ritter 1999, 2003). On the other hand, choppers, knives, and other cutting tools may have been used in the dismemberment and filleting of marine mammals (Oviedo-García, personal communication 2013), while elements such as cobbles may have been used to dismember the animal by breaking the articulations (Lyman 1992), as well as breaking the bones to extract marrow.

In other regions, harpoons were the tools most used to capture marine mammals (Hildebrandt & Jones 1992), either on land or from boats and rafts (Ritter 1999). On the peninsula, the presence of harpoons points is rare, and they are limited to a few sites in the central region, including Isla Cedros (Des Lauriers 2006; Ritter 1998a, 1999). They are not manifested as a tradition that diffused throughout Baja California, and they seem most likely to be absent in the northern region, or at least that they have not been found yet. Nonetheless, many historic sources make reference to the use of harpoons in the southern region (Laylander 2005). It is speculated that several projectile points found in the central area of Baja California may have functioned as harpoons inset (Ritter and Burcell 1998), perhaps to obtain a more effective weapon that could be used in the hunting of marine mammals. However, this only seems to have happened at coastal locations in the west-central region of the peninsula (Ritter 2006b).

Finally, watercraft served to transport people and goods, but they were also an element used in hunting marine mammals (Arnold 1992; Hildebrandt & Jones 1992), and their existence is reported from at least 8000 B.P. on the California coast (Cassidy et al. 2004). One of the well-documented examples of the presence of watercraft is found on Isla Cedros. Des Lauriers (2005) reported a type of watercraft effective for maritime activities, with the ability to cross the channel between the island and the peninsula. However, it could also have assisted in the hunting of pinnipeds and smaller cetaceans through the exploitation of zones far out from the beach or difficult to access by land (Des Lauriers 2006).

Marine Mammals As Raw Material

Marine mammals definitely served a dietary function among the groups settled at the coastal sites of the Baja California peninsula, as is manifest in the numerous remains with burning and with butchering cuts from dismemberment and filleting. They also served possible ritual functions, in accompanying burials both as modified and as unmodified bones (Guía-Ramírez 2008b; Oviedo-García & Guía-Ramírez 2005). Furthermore, fat, pelts, and intestines were harvested but archaeological evidence of them does not survive, and it only possible to affirm their use through inferences from ethnohistorical sources.

Use as raw material was also widely diffused throughout the region. Cetacean bones were mostly used for the manufacturing of various objects. Several tools for detaching abalone, chitons, and some limpets from rocky substrates were certainly made from the bones of large cetaceans (Guía-Ramírez 2006; Ritter 1999). Likewise, some mortuary elements that had been formed into oars or paddles and were deposited on burials were obtained from flat bones such as ribs and/or mandibles of the same group (Oviedo-García & Guía-Ramírez 2005). Harpoons of diverse forms and sizes were produced as an adaptation to a maritime lifeway, for fishing and the hunting of turtles and marine mammals; their characteristics are consistent with whale bone (Des Lauriers 2006; Ritter 1999, 2002, 2006a).

On the peninsula, the use of cetacean bones as raw material goes back to the early Holocene. At the site of Abrigo de los Escorpiones, in the northwest region, a large object from a whale mandible, presumably from a gray whale (*Eschrichtius robustus*), was reported; its age was established as 8100 ±40 B.P. according to radiocarbon results based on charcoal directly associated with the element (Bryan & Gruhn 2005). Other finds indicate that this practice continued throughout the Holocene (Oviedo-García & Guía-Ramírez 2005) but diminished considerably toward the Late Prehistoric period (Guía-Ramírez 2006).

TEMPORAL VARIATION IN THE EXPLOITATION OF MARINE MAMMALS

Undoubtedly, the dependence on coastal resources and the manner of exploiting them changed throughout the peninsula. The types of resources and their exploitation changed along north/south axes, between the east and west coasts, and undeniably also through time.

Although very old sites – late Pleistocene and early Holocene – have been recorded across the peninsula, they are rare. This situation is explained by the history of variation in sea level that resulted in early coastal sites becoming submerged (Davis 2006). Nonetheless, as has been noted above, marine mammal remains are present in archaeological sites since the early Holocene. How the exploitation of this resource varied through time is still not clear for the greater part of the peninsula. However, in the southern part of the peninsula it appears that the use of marine resources increased up to the late Holocene (Laylander & Moore 2006). This was the period when the specialized exploitation of dolphins was established at site Las Tinas 3 in Baja California Sur (Fujita 2006; Porcasi & Fujita 2000).

In the northern part of the peninsula, primarily on the west coast, a temporal variation in the use of marine mammals is seen (Guía-Ramírez 2010b). However, much is left to explore, because of many sites have not been studied or do not have radiometric records to establish dates. Table 1 is a summary of the representation of sea mammals at seven sites in the northwest of Baja California -three from the Archaic and four from the Late Prehistoric period- at which the location and the period are well defined. There exists one other site that can be assigned to a period by technology and similar methods but whose age cannot be certainly established; consequently, it was not considered in this chapter. There is evidence that the use of these animals was frequent and repeated for sites and strata assigned to the Archaic period (7,500 to 1,500 BP). In this period, human groups adopted a closer relationship with marine resources, in response to climatic changes prevailing in the region after leaving a cooler and more humid climate, when the early inhabitants had preferred to live near lakes and other bodies of fresh water, subsisting on terrestrial mammals and aquatic birds. With a more arid climate and the desiccation of many of the bodies of fresh water, human groups modified their lifeways, incorporating marine resources as a key component of the subsistence economy. This was when shell middens appeared in most coastal zones that were occupied and frequented up to European contact (Davis 2006). It was precisely during this period that marine mammals may have been converted into an economically valuable resource. However, it remains unclear whether a specific hunting pattern existed, since elements needed to clarify this activity are lacking.

Table 1. Remains of marine mammals in seven archaeological sites in the northwest region of Baja California

	NISP						
Taxa	Archaic sites			Late Prehistoric sites			
	L20	EM	B	L20	LP	J	B
Enhydra lutris	9	18	19	4	1	--	24
Otariidae	2	--	--	--	--	--	--
Zalophus californianus	14	7	32	14	1	--	14
Delphinidae	2	--	--	--	--	--	--
Cetaceans	2	7	4	5	--	--	2

On the other hand, already in the Late Prehistoric period (1,500 to 410 BP) the remains of mammals appear to have decreased in archaeological contexts, and perhaps even to have disappeared from them (Guía-Ramírez 2008c, 2010), which denoted a departure from the economic systems that had prevailed during the Archaic period. Around 1500 B.P., there was a radical change in the means of subsistence that some investigators relate to the ENSO (El Niño-Southern Oscillation) phenomenon (Colten & Arnold 1998). This is characterized by a rise in surface ocean temperature, a decrease in the upwelling of cold water that is rich in nutrients, and consequently a decrease in primary productivity; combined with other abnormal conditions, this result in changes in marine ecosystems. These "abnormal" conditions, such as the rise in surface water temperature and changes in the availability and abundance of food resources, seem to be factors to which marine mammals, especially pinnipeds, are sensitive (Colten 1993; Colten & Arnold 1998; Porcasi & Fujita 2000). Undoubtedly they affect all marine resources. Thus, in the face of adverse conditions, groups initiated the exploitation of acorns, other seeds, and fruits, and intensified the exploitation of inland resources, implementing economic diversification. Through economic reorganization, in which semipermanent settlements were replaced with seasonal mobility – involving trips to the coast in the winter and to the mountains in the summer – and with the exploitation of interior resources (Guía-Ramírez & Oviedo-García 2009). The economic value of marine mammals diminished, because people wanted a method of subsistence with less expenditure of energy on maritime activities. Such a method was the collection of mollusks, sea urchins, and crabs, as well as fish species, mainly from the kelp beds that are abundant in the north of the region.

Another factor that may have influenced the exploitation of new resources, and with it the decline in the use of marine mammals, was the introduction of the bow and arrow, which permitted an increase in the effectiveness of hunting terrestrial mammals such as deer, mountain sheep, rabbits, and hares (Ritter 2006a). The introduction of the bow and arrow took place during the final portion of the Late Prehistoric period (Laylander 1999).

What has been observed in the preceding paragraphs explains what is now known about the reasons for the variation in marine mammal remains in the archaeological contexts of the northwest part of the Baja California peninsula. However, it does not reflect what happened in the rest of the territory, where future studies may be able to shed light on this theme.

OTHER ASPECTS OF MARINE MAMMALS

So far, the presence of marine mammal remains seems to be limited to coastal sites. Even when other products of marine origin have been detected in the interior of the peninsula (Fujita 2006; Guía-Ramírez & Oviedo-García 2009; Gutierrez & Hyland 2002; Oviedo-García 2006; Ritter 2006a, 2006c), even in the higher mountains at several sites found more than 100 km distant from the coast, such as in the Sierra Juárez (Oviedo-García & Guía-Ramírez 2009), no archaeological evidence for the transporting of any products related to these mammals has been recorded.

Nonetheless, the use of marine mammal hides in general and of sea lion bladders for carrying water are described in ethnohistoric sources. According to Des Lauriers (2006), Ulloa and Preciado – early European explorers of the peninsula – saw and described the intensive processing of hides on Isla Cedros. Indeed, it is suggested that production exceeded the islanders' needs, and consequently that the excess hides may have been used for exchange with mainland groups. The availability of water is one of the resources limiting the establishment of human settlements, mainly in desert areas. In some cases, human groups use the internal organs such as bladders of marine animals to contain or carry this element (González-Vázquez, 2000). Del Barco (1988) described the use of sea lion bladders to transport water. Perishable products from marine mammals may have been used and carrying between sites, including from the coast to the interior, but it remains difficult to establish this conclusively because it is not detectable archaeologically.

There is indirect evidence concerning knowledge about marine mammals and perhaps the importance that they had for native groups who visited interior valleys and mountains, in the form of the diverse rock art that pictures animals such as whales and possibly sea lions (Crosby 1997). Those may also have been related to successful hunting by means of rituals (Ritter 2006a).

CONCLUSION

According to archaeological studies completed so far, marine mammals do not seem to have dominated archaeological contexts, but their use by human groups may be underestimated, above all if the quantity of protein and fat that an animal could supply is not considered in relation to the quantity of wastes that remain, particularly when they are compared with mollusks. Despite this, it appears that there was no intensive or specialized exploitation of marine mammals in most of the peninsula, but rather a moderate exploitation that was based on limited hunting and the use of beached animals. However, those places should not be neglected where there was intensive exploitation in procuring and processing marine resources and where marine mammals were a highly significant resource, as in the case of the islands or the Cape region, which by their geographical characteristics may represent insular systems.

To conclude, the Baja California peninsula represents one of the most enigmatic regions of Mexico, with diverse mosaics, both cultural and environmental. Until a few years ago, very little was known concerning the cultural development of the human groups that occupied it. Further studies will make possible a better interpretation of the interrelation between marine resources and the peninsula's first inhabitants.

REFERENCES

Arnold, J. E. (1992). Complex Hunter-Gatherer-Fishers of Prehistoric California: Chiefs, Specialists, and Maritime Adaptations of the Channel Islands. *American Antiquity 57,* 60-84.

Del Barco, M. (1988). *Historia Natural y Crónica de la Antigua California.* Mexico City: Universidad Nacional Autónoma de México, Instituto de Investigaciones Históricas.

Bendímez Patterson., J., M. A. Téllez & Serrano, J. (1993). Excavaciones arqueológicas en el poblado de Bahía de los Ángeles. *Estudios Fronterizos 31/32*, 175-216.

Bryan, A. & Gruhn, R. (2005). *Informe sumario de la temporada cuarta de excavaciones en el Abrigo de Los Escorpiones, Baja California, Informe Técnico*. Mexicali, Mexico: Centro INAH-Baja California.

Cassidy, J. L. M. Raab & Kononenko, N. A. (2004). Boats, Bones, and Biface Bias: The Early Holocene Mariners of Eel Point, San Clemente Island, California. *American Antiquity 9*, 109-130.

Celis, A. K. (2011). *Explotación de recursos costeros y condiciones ambientales en el delta del Río Colorado durante la Prehistoria.* Unpublished Master's thesis, presented at the Universidad Autónoma de Baja California, Ensenada, Mexico.

Colten, R. H. (1995). Faunal Exploitation during the Middle to Late Period Transition on Santa Cruz Island, California. *Journal of California and Great Basin Anthropology 17(1)*, 93-120.

Colten, R. H., & Arnold, J. E. (1998). Prehistoric Marine Mammal Exploitation on California's Northern Channel Islands. *American Antiquity 63,* 679-701.

Crosby, H. (1997). *The Cave Paintings of Baja California: Discovering the Great Murals of an Unknown People*. San Diego, USA: Sunbelt Publications.

Davis, L. (2006). Baja California´s Paleoenvironmental Context. In D. Laylander & J. Moore (Eds.), *The Prehistory of Baja California: Advances in the Archaeology of the Forgotten Peninsula* (pp. 14-23). Gainesville, USA: University Press of Florida.

Des Lauriers, M. R. (2005). The Watercraft of Isla Cedros, Baja California: Variability and Capabilities of Indigenous Seafaring Technology along the Pacific Coast of North America. *American Antiquity 70(2)*, 342-360.

Des Lauriers, M. R. (2006). Terminal Pleistocene and Early Holocene Occupations of Isla de Cedros, Baja California, Mexico. *Journal of Island & Coastal Archaeology 1*, 255-270.

Des Lauriers, M. R. (2006). Isla Cedros. In D. Laylander & J. Moore (Eds.), *The Prehistory of Baja California: Advances in the Archaeology of the Forgotten Peninsula* (pp. 153-166). Gainesville, USA: University Press of Florida.

Drakic, D., L. Delgado & Cuadra, O. (2005). *Salvamento Arqueológico Costa Azul-Lote 20*, Informe Técnico. Mexicali, Mexico: Centro INAH-Baja California.

Drakic, D., L. Delgado & Cuadra, O. (2006). *Salvamento arqueológico Corredor Tijuana-Rosarito*, Informe Técnico. Mexicali, Mexico: Centro INAH-Baja California.

Fujita, H. (2006). The Cape Region. In D. Laylander & J. Moore (Eds.). *The Prehistory of Baja California: Advances in the Archaeology of the Forgotten Peninsula* (pp. 82-98). Gainesville, USA: University Press of Florida.

Fujita, H. & Guía-Ramírez, A. (2011). Restos de fauna en varias localidades de la Isla Espíritu Santo, B.C.S. Encuentro Binacional Balances y Perspectivas. Mexicali, Mexico: Centro INAH-Baja California.

Fujita, H. & Poyatos, G. P. (1998). Settlement Patterns on Espíritu Santo Island, Baja California Sur. *Pacific Coast Archaeological Society Quarterly 34(4)*, 67-105.

Gifford-Gonzalez, D., Newsome, S., Koch, P., Guilderson, T., Snodgrass, J. & Burton, R. (2004). Archaeofaunal Insights on Pinniped-Human Interactions in the Northeastern Pacific. In G. Monks (Ed.). *The Exploitation and Cultural Importance of Sea Mammal*s (pp. 19-38). Oxford, UK: Oxbow Books.

Glassow, M. A. (1997). Middle Holocene Cultural Development in the Central Santa Barbara Channel Region. In J. M. Erlandson & M. A. Glassow (Eds.). *Archaeology of the California Coast during the Middle Holocene* (pp. 73-90). Los Angeles, USA: Institute of Archaeology, University of California.

González-Vázquez, A. (2000). Aprovechamiento de los recursos acuáticos en California y Alaska: cazadores-recolectores en las fuentes escritas de los Siglos XVI, XVII y XVIII. *Nivel Cero 8*, 77-93.

Guía-Ramírez, A. (2005). Diversidad faunística en los concheros de Costa Azul: la vida junto a las costas. VI Encuentro Binacional: Balances y Perspectivas 2005. Mexicali, Mexico: Centro INAH- Baja California.

Guía-Ramírez, A. (2006). El hueso como materia prima entre los antiguos californios, VII Encuentro Binacional Balances y Perspectivas 2006. Mexicali, Mexico: Centro INAH Baja California.

Guía-Ramírez, A. (2007). Conchas, concheros y las aguas del Golfo de California: Análisis arqueozoológico de restos biológicos del área de Mexicali. In Porcayo, A. (Ed.). *Proyecto registro y rescate de sitios arqueológicos de Baja California: Fase Municipio de Mexicali-temporada 2007*, Informe Técnico. Mexicali, Mexico: Centro INAH-Baja California.

Guía-Ramírez, A. (2008a). Indigenous Fishing Practices in Ancient California. 42nd Annual meeting of the Society for California Archaeology, Burbank.

Guía-Ramírez, A. (2008b). Fauna asociada a un entierro múltiple en el Desierto Central, B.C. IX Encuentro Binacional, Balances y Perspectivas, Tecate, Mexico.

Guía-Ramírez, A. (2008c). Variación en el tiempo: Los mamíferos marinos en la arqueología de Baja California. Reunión Internacional de Mamíferos Marinos, Ensenada, Mexico.

Guía-Ramírez, A. (2009). Bivalves and Gastropods in the Gulf of California: Taking Stock on the Shorelines. *Proceedings of the Society for California Archaeology* 23.

Guía-Ramírez, A. (2010a). Una retrospectiva de los mamíferos marinos en la arqueología de Baja California. *Memorias de la 32 Reunión Internacional para el estudio de los mamíferos marinos*. Veracruz, México.

Guía-Ramírez, A. (2010b). Estudio comparativo de arqueofaunas al norte de BC: un enfoque temporal, *Informe final: Análisis de materiales arqueozoológicos en Baja California*. Mexicali, Mexico: Centro INAH B.C.

Guía-Ramírez, A. & Oviedo-García, F. (2009). Los antiguos californios y sus viajes de las costas a la montaña. *Camino Real Misionero de las California 6*, 16-21.

Gutiérrez, M. L. & Hyland, J. (2002). *Arqueología de la Sierra de San Francisco: Dos décadas de investigación del fenómeno Gran Mural.* Mexico City: Colección Científica 433INAH.

Hildebrandt, W. R. & Jones, T. (1992) Evolution of Marine Mammal Hunting: A View from the California and Oregon Coasts. *Journal of Anthropological Archaeology 11*, 306-401.

Jones, T. L. & Hildebrandt, W. R. (1995). Reasserting a Prehistoric Tragedy of the Commons: Reply to Lyman. *Journal of Anthropological Archaeology 14*, 78-98.

Laylander, D. (2005). Large Projectiles and the Cultural Distinction of Southern Baja California: A Reexamination. *Pacific Coast Archaeological Society Quarterly 39(2-3)*, 11-21.

Laylander, D. (1999). Una exploración de las adaptaciones culturales prehistóricas de Baja California. In Bendímez Patterson, J. (Ed.). *Antología de la arqueología de Baja California* (pp. 41-45). Mexicali, Mexico: Centro INAH-Baja California.

Laylander, D. (2006). Issues in Baja California Prehistory. In D. Laylander, & J. Moore (Eds.). *The Prehistory of Baja California: Advances in the Archaeology of the Forgotten Peninsula* (pp. 1-13). Gainesville, USA: University Press of Florida.

Laylander, D. & Moore, J. (Eds.), (2006). *The Prehistory of Baja California: Advances in the Archaeology of the Forgotten Peninsula.* Gainesville, USA: University Press of Florida.

Lazcano, C. S. (2000). *Pa Tai: La historia olvidada de Ensenada.* Colección de documentos sobre la historia y la geografía del municipio de Ensenada No. 2. Ensenada, Mexico.

Lyman, R. L. (1989). Seal and Sea Lion Hunting: A Zooarchaeological Study from the Southern Northwest Coast of North America. *Journal of Anthropological Archaeology* 8(1):68-99.

Lyman, R. L. (1992). Prehistoric Seal and Sea-lion Butchering on the Southern Northwest Coast. *American Antiquity* 57(2):246-261.

Lyman, R. L. (1995). On the Evolution of Marine Mammal Hunting on the West Coast of North America. *Journal of Anthropological Archaeology* 14:45-77.

Maya, Y., F. R. Venegas & Manríquez, F. J. (2011). Geografía de suelos regional: Península de Baja California. In P. Krasilnikov, F. Jiménez, T. Reyna & García, N. (Eds.). *Geografía de suelos de México*. Mexico City: UNAM Facultad de Ciencias.

McCawley, W. (1996). *The First Angelinos: The Gabrielino Indians of Los Angeles*. Banning (California), USA: Malki Museum Press.

Medrano, L. E., P. Recagno, M. J. Vázquez & Rosales Nanduca, H. (2007). *Los mamíferos marinos ante el cambio ambiental en el Pacifico Tropical Mexicano*. Mexico City: CONABIO, Biodiversitas.

Ogden, A. (1941). *The California Sea Otter Trade, 1784-1848.* Berkeley, USA: University of California Press.

Oviedo-García, F. (2008). *Informe del proyecto de salvamento "Estudio para la liberación del gasoducto Costa Azul-El Florido, B.C"*, Informe Técnico. Mexicali, Mexico: Centro INAH-Baja California.

Oviedo-García, F. & Guía-Ramírez, A. (2004). Primeras interpretaciones del rescate arqueológico en el conchero El Morro, Rosarito, Baja California. *Memorias balances y perspectivas 2002-2004, V encuentro binacional, Las costas prehispánicas e hispánicas de Baja California*. Mexicali, Mexico: Centro INAH-Baja California.

Oviedo-García, F. & Guía-Ramírez, A. (2009). Characteristics of a Shellfish Collecting Camp at Costa Azul, Baja California. *Proceedings of the Society for California Archaeology* 23.

Porcasi, J. F. & Fujita, H. (2000). The Dolphin Hunters: A Specialized Prehistoric Maritime Adaptation in the Southern California Channel Islands and Baja California. *American Antiquity 65(3)*, 543-566.

Porcasi, J. F., T. L. Jones & Raab, L. M. (2000). Trans-Holocene Marine Mammal Exploitation on San Clemente Island, California: A Tragedy of the Commons Revisited. *Journal of Anthropological Archaeology 19*, 200-220.

Porcayo, A. (2008). *Proyecto registro y rescate de sitios arqueológicos de Baja California: fase Municipio de Mexicali*, Informe Técnico parcial de la primera temporada y propuesta para trabajo de campo 2007. Mexicali, Mexico: Centro INAH-Baja California.

Renker, A. M. & Gunther, E. (1990). Makah. In Suttles, W. (Ed.). *Northwest Coast, Handbook of North American Indians, vol 7* (pp. 422-430). Washington, D.C., USA: Smithsonian Institution.

Ritter, E. (1998). Investigations of Prehistoric Behavioral Ecology and Culture Change within the Bahía de Los Angeles Region, Baja California. *Pacific Coast Archaeological Society Quarterly 34(3)*, 9-43.

Ritter, E. (1999). *Investigaciones arqueológicas en Laguna Guerrero Negro, Baja California (Fase 1)*, Informe Técnico. Mexicali, Mexico: Centro INAH-Baja California.

Ritter, E. (2002). The Archaeology of the Three Sisters Lagoons, Baja California. In *Memorias de Balances y Perspectivas de la Antropología e Historia de Baja California 2002-2004*, pp. 61-74. Mexicali, Mexico: Centro INAH-Baja California.

Ritter, E. (2006a). South-Central Baja California, In D. Laylander, & J. Moore (Eds.), *The Prehistory of Baja California: Advances in the Archaeology of the Forgotten Peninsula*, pp. 99-116. Gainesville, USA: University Press of Florida.

Ritter, E. (2006b). The Vizcaíno Desert. In D. Laylander & Moore, J. (Eds.), *The Prehistory of Baja California: Advances in the Archaeology of the Forgotten Peninsula* (pp. 135- 152). Gainesville, USA: University Press of Florida.

Ritter, E. (2006c). Bahía de Los Ángeles. In D. Laylander, & Moore, J. (Eds.). *The Prehistory of Baja California: Advances in the Archaeology of the Forgotten Peninsula* (pp. 167- 178). Gainesville, USA: University Press of Florida.

Ritter, E. (2012). Comprehending the Prehistory of Laguna Manuela, Baja California: A Summary. *Proceedings of the Society for California Archaeology 26*, 51-70.

Ritter, E. & Burcell, J. (1998). Projectile Points from the Three Sisters´ Lagoons of West Central Baja California. *Pacific Coast Archaeological Society Quarterly 34(4)*, 29-66.

Roman, D. V. & Des Lauriers, M. (2010). Debating Changing Late Holocene Subsistence Strategies: Evidence from Analogies between Central Baja California and the Santa Barbara Basin Region (Channel Islands). *Proceedings of the Society for California Archaeology* 24.

Sanchez, S. J. (2009). Reporte cuantitativo de la arqueofauna de vertebrados del salvamento Cañada del Águila, Sitio La Punta, Baja California, México. In D. Drakic (Ed.). *Salvamento arqueológico Cañada del Águila: Segunda temporada*, Informe Técnico. Mexicali, Mexico: Centro INAH-Baja California.

Sheehan, G.W. (1985). Whaling as an Organizing Focus in Northwestern Alaskan Eskimo Societies. In T. D. Price & J. A Brown (Eds.). *Prehistoric Hunter-Gatherers: The Emergence of Cultural Complexity* (pp. 123-154). San Diego, USA: Academic Press.

Torres, A., C. Esquivel & Ceballos, G. (1995). Diversidad y conservación de los mamíferos marinos de México. *Revista Mexicana de Mastozoología 1*, 22-43.

Vargas, J. (2007). The California Gray Whale: Its Legal Regime under Mexican Law. *Ocean and Coastal Law Journal 12(2),* 213-244.

Wake, T. (1999). Temporal Variation in Vertebrate Archaeofaunas from Camp Pendleton Marine Corps Base, San Diego County, California. *Pacific Coast Archaeological Society Quarterly 5(4)*,45-64.

Walker, P. L., J. K. Douglas, T. L. Jones & DeLong, R. (2000). Archaeological Investigations at the Point Bennett Pinniped Rookery on San Miguel Island. *Proceedings of the Fifth California Islands*

Symposium, March 29-April 1 (pp. 628-632). Santa Barbara Natural History Museum.

Zepeda-Gutiérrez, J. (1986). Reseña histórica de la cacería de nutrias de mar en Baja California, México. *Memoria del Primer Simposio Nacional sobre el desarrollo de las investigaciones oceanográficas en México*. Mexico City.

In: Neotropical and Caribbean Aquatic Mammals ISBN: 978-1-63321-306-7
Editors: Muñoz, Götz and Ramos Roca © 2014 Nova Science Publishers, Inc.

Chapter 2

THE ARCHAEOLOGY OF THE INTERACTION BETWEEN MARINE MAMMALS AND HUMANS IN CENTRAL AND SOUTHEASTERN MESOAMERICA

Christopher Markus Götz[1*], Raúl Valadez Azúa[2†] and Bernardo Rodríguez Galicia[2‡]

[1]Facultad de Ciencias Antropológicas,
Universidad Autónoma de Yucatán, Mexico
[2]Instituto de Investigaciones Antropológicas,
Universidad Nacional Autónoma de México, Mexico

ABSTRACT

The chapter describes the zooarchaeological data of aquatic mammal exploitation in central, but mainly southeastern Mesoamerica. The area, which housed the great pre-Columbian civilizations like the Maya or Teotihuacans, indicates a relatively low level of exploitation of aquatic animals. From the Yucatan peninsula, a set of 19 sites is described, where either zooarchaeological material (e.g. bone or teeth remains) or iconographic evidence of aquatic mammals were found. Among the

* cgotz@correo.uady.mx
† raul_valadez@hotmail.com
‡ sanber65@hotmail.com

Maya, the most frequently exploited aquatic mammal was the manatee (*Trichechus manatus*), although several sites of the northern peninsula's coast also yielded bone and teeth fragments of the west Indian Seal (*Monachus tropicalis*). According to ethnohistorical accounts, aquatic mammals were considered important in Ancient Mesoamerica, since not only their meat and fat was used, but the bones also served as raw material for sculptures, and small figurines representing seals or even whales were sometimes crafted out of clay or stone. Yet, it is believed that the Ancient Mesoamericans did not frequently hunt mammals in the coastal waters, since the agricultural *milpa* system offered sufficient access to a diet based on vegetables and fauna.

Introduction

Skeletal materials of marine mammals are a relatively rare phenomenon at pre-Hispanic archaeological sites from central and southeastern Mesoamerica. Most finds are limited to only a few specimens mixed with the general archaeological matrix together with a high amount and variety of other faunal remains, while on the other hand there is very little information from pre-Columbian iconography and codices, or even colonial manuscripts that would allow us to understand the vision and concepts that pre-Hispanic Indians had of these animals. In spite of such scarceness, it is possible to retrieve some interpretations based on data derived from archaeozoological remains, historical sources and iconography about the use-patterns and meaning of aquatic mammals for the indigenous population in central Mexico and the Maya area.

In the present chapter, we intend to reach such interpretations, by exploring the level of contact between sea mammals and prehispanic Indians of Mesoamerica synthesizing several aspects. After a brief zoological characterization of the common aquatic mammals found in and around Mesoamerica, we will scrutinize the hunting techniques and exploitation patterns of marine mammals by the pre-Columbian population as described in the historic sources written at the time of the conquest or shortly thereafter.

The description of the historic situation of the sea mammals and their relation to humans (or vice versa) is followed by an analysis of the archaeological data that outlines the eventual exploitation pattern of those animals, treating where possible the role of aquatic mammals in Meso-american iconography through time. The archaeological and iconographical portrayals are headed by a brief description about the earliest paleontological finds of aquatic mammals in the area. At the end of the chapter, serving as a discussion, we present a general panorama of the exploitation of aquatic mammals in ancient Mesoamerica, trying to combine the historic information with the data from archaeological contexts. One of the most prevalent aquatic mammals we treat in this chapter is the manatee, whose exploitation in the Maya area was studied in detail before; we are thus tying our chapter up to McKillop's (1985) excellent study about this taxon, that was published almost 30 years ago.

MESOAMERICA'S AQUATIC MAMMALS - A VIEW TOWARDS THE OPEN SEA, THE RIVERS, ESTUARIES AND MANGROVES

The waters in and around Mesoamerica are the home of a relatively high number of aquatic mammals (see definition in Muñoz, Götz & Ramos, this volume), some occurring in rivers, others in the estuaries surrounded by mangroves, and still others more in the Gulf of Mexico, the Caribbean Sea and the Pacific Ocean. Reid (1997) lists a total of 31 species of aquatic mammals for Central America and Southeastern Mexico, many of which live their whole life in the ocean.

These 31 species belong to two orders -Sirenidae (manatees) and Cetacea (whales and dolphins)- and 6 families -Trichechidae (New World manatees), Delphinidae (dophins), Physeteridae (sperm whales), Ziphiidae (beaked whales), Balaenopteridae (rorqual whales), and Baleanidae (right and bowhead whales). To this list, a 32nd -now extinct- species of the Carnivora order, Phocidae family and Monachinae subfamily must be added, the West Indian Seal (Adam 2004; Hairr 2011). As will be seen by the following parts of the chapter, the vast majority of archaeological, iconographic or historical data from Mesoamerica refer the prehispanic exploitation of the manatee and the seal. Archaeological finds of whalebones or parts thereof are very limited at Mesoamerican sites, either inland or coastal (see Corona-Martínez, Carmona Macías, and Arroyo-Cabrales, this volume). It was therefore decided to only describe the manatee and seal in detail, while the characteristics of the other aquatic mammals are resumed in a table.

The West Indian manatee (*Trichechus manatus*; Figure 1), an animal belonging together with its distant proboscidean relatives to the clade

Tethytheria, is only found at the east coast of Mesoamerica, corresponding to the shores of the Gulf of Mexico and the Caribbean (Reid 1997:276-277).

Figure 1. Drawing of a manatee (*Trichechus manatus*), used with permission of the CONABIO; Copyright: Marco Antonio Pineda Maldonado/CONABIO.

According to Reid (op. cit.), adult manatees, whose name in Maya was teek, range between 100 and 300 kg, although some sources even register weights up to 750 kg (McKillop 1985:339; Spalding 2004:55). The manatee is a silent and relatively slow-moving herbivore that browses sea grasses and other aquatic vegetation in shallow waters (Reid 1997:276), which might be the reason why it is called *vaca marina* (sea cow) in Spanish.

The extinct West Indian Seal (*Monachus tropicalis* (Gray 1850); Figure 2) (Davies 2008; Adam 2004) belonged to the family of the true seals (Mangel 2003:3) and is recorded for the Gulf of Mexico and Caribbean Sea. This animal, which was called *tsula* in Maya (Andrews 1984:4), reached up to 135 kg (Davies 2008) and fed on fish and crustaceans (Adam and García 2003:305; Adam 2004:3) that were probably caught near the reefs and around remote islands far from the mainland (Adam & García 2003:305; Hairr 2011; see also Götz 2012a).

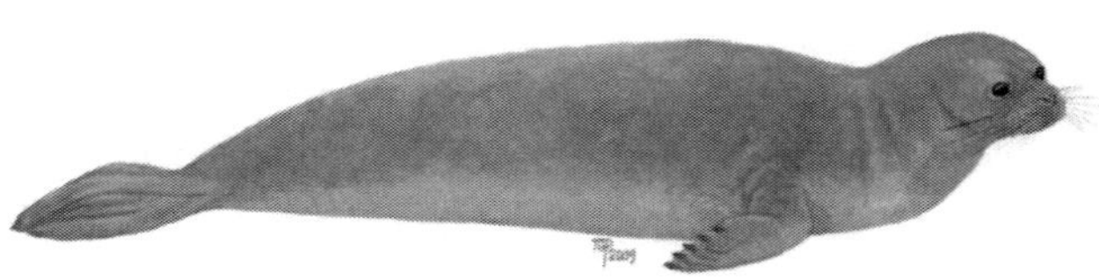

Figure 2. Drawing of a West Indian seal (*Monachus tropicalis*), used with permission of the CONABIO; Copyright: Marco Antonio Pineda Maldonado/CONABIO.

The group of dolphins and whales that approaches Mesoamerica's coastal waters is composed by 29 species, whose general characteristics are resumed in Figure 3.

Taxonomic group / family	Common and scientific name	Approximate weight	Preferred food	Habitat
Ocean Dolphins (Delphinidae)	Rough-toothed dolphin (*Steno bredanensis*)	120 - 160 kg	squid, octopus, and large fish	deep, offshore water; occasionally enters shallow coastal waters
	Tucuxi (*Sotalia fluviatilis*)	35 - 50 kg	fish and crustaceans	coastal waters, major rivers, estuaries, inshore waters (fresh and salt water)
	Pantropical spotted dolphin (*Stenella attenuata*)	100 - 120 kg	flying fish and squid, among others	offshore waters, but also coastal regions with deep water
	Atlantic spotted dolphin (*Stenella frontalis*)	100 - 140 kg	invertebrates and small schooling fish (Allen 2001)	offshore and coastal waters
	Spinner dolphin (*Stenella longirostris*)	70 - 95 kg	small fish and squid	deep, offshore water, sometimes inshore waters near islands
	Clymene dolphin (*Stenella clymene*)	75 - 85 kg	small fish and squid (Benton 2000)	deep offshore waters
	Striped dolphin (*Stenella coeruleoalba*)	100 - 150 kg	small fish and squid	deep and offshore waters
	Common dolphin (*Delphinus delphis*)	70 - 135 kg	small fish and squid	deep and offshore waters
	Bottlenose dolphin (*Turiopsis truncatus*)	200 - 650 kg	bottom-dwelling fishes and invertebrates	coastal waters, estuaries, major rivers, offshore waters
	Fraser's dolphin (*Lagenodelphis hosei*)	100 - 210 kg	fish, squid and crustaceans	deep and offshore waters
	Risso's dolphin (*Grampus griseus*)	300 - 500 kg	squid and crustaceans	deep, offshore waters
	Melon-headed whale (*Peponocephala electra*)	150 - 275 kg	medium sized fish and squid	offshore waters
	Pygmy killer whale (*Feresa attenuata*)	150 - 225 kg	squid and fish	deep and offshore waters
	False killer whale (*Pseudorca crassidens*)	1,000 - 2,500 kg	large fish and squid	deep and offshore waters
	Killer Whale (*Orcinus orca*)	4,000 - 8,000 kg	marine mammals, fish, and other vertebrates	cold, inshore waters

Figure 3. (Continued on next page)

Taxonomic group / family	Common and scientific name	Approximate weight	Preferred food	Habitat
	Short-Finned Pilot Whale (*Globicephala macrorhynchus*)	1,000 - 3,000 kg	squid and fish	deep and offshore waters
Sperm Whales (Physeteridae)	Great Sperm Whale (*Physeter catodon*)	22,000 - 57,000 kg	mainly deep-water squid and octopus	deep offshore waters
	Pygmy Sperm Whale (*Kogia breviceps*)	350 - 400 kg	mainly deep-water fish, squid and octopus	deep offshore waters
	Dwarf Sperm Whale (*Kogia simus*)	150 - 210 kg	mainly deep-water squid and octopus	deep offshore waters
Beaked Whales (Ziphiidae)	Cuvier's Beaked Whale (*Ziphius cavirostris*)	3,000 - 4,000 kg	deep-water squid and fish	deep offshore waters
	Blaineville's Beaked Whale (*Mesoplodon densirostris*)	470 - 1,030 kg	deep-water squid and fish	deep offshore waters
	Gervais' Beaked Whale (*Mesoplodon europeus*)	560 - 1,200 kg	squid	deep offshore waters
Rorqual Whales (Balaenopteridae)	Minke Whale (*Balaenoptera acutorostrata*)	6,000 - 14,000 kg	krill and schooling fish	coastal and inshore waters, rarely offshore
	Bryde's Whale (*Balaenoptera edeni*)	13,000 - 25,000 kg	small fish	coastal and offshore waters
	Sei Whale (*Balaenoptera borealis*)	14,000 - 30,000 kg	squid, krill and small fish	deep offshore waters
	Fin Whale (*Balaenoptera physalis*)	40,000 - 75,000 kg	squid, krill and small schooling fish	deep offshore waters, rarely coastal waters
	Blue Whale (*Balaenoptera musculus*)	90,000 - 190,000 kg	krill and other crustaceans	offshore waters
	Humpback Whale (*Megaptera novaeangliae*)	24,000 - 35,000 kg	krill and small fish	shallow coastal waters
Right and Bowhaed Whales (Baleanidae)	Northern Right Whale (*Eubalaena glacialis*)	60,000 - 160,000 kg	squid and krill (Scott 2002)	shallow coastal waters

Figure 3. General characteristics of the dolphin and whale species of Mesoamerica's coast. All data shown are based on Reid (1997:287-300), unless otherwise noted.

"With mustard, their flesh tastes almost like beef[1]*"*

Exploitation Patterns and Hunting Techniques of Aquatic Mammals Described in the Historic Sources

Soon after the Spanish conquest, Spanish missionaries and other erudites wrote detailed chronicles about the New World. Among the treaties concerning the religious conditions of the Indian population and the history of the Spanish conquest are also several references to the *natural* world, the environment and the animals, many so different from Europe, that the Spaniards came across upon their arrival. The sea cow or manatee for example, which resembled the morphological and biological properties of fish and cow, was completely new to Spanish eyes, as can be seen in the accounts shown or referred to below. On the other hand, several aquatic mammals that were observed and described by the Spanish chroniclers, such as whales, dolphins and seals, were well known in Europe, as they were hunted since Medieval times in Spain and other countries with access to the sea (see, among others, Boyle 2005; Sabin 2005; Stora 2002; Stora & Lougras 2005).

The early historic reports about aquatic mammals describe the properties of those taxa in accounts about the use the Indian population made of them, which then depended on the area and culture. In this sense, descriptions about whales and dolphins are limited to chronicles about the West Indies and Florida, because the peoples that lived there exploited cetaceans. Instead, descriptions about the Mesoamerican cultures at the eve of the Conquest and the fauna those people used, include only the manatee as an important aquatic mammal for Indian groups of the Maya area[2]. Historic sources of Mesoamerica's center, written and gathered by Indian annalists or informants, such as the *Crónica Mexicana* by Hernando de Alvarado Tezozomoc (1987) or the *Historia General de las Cosas de la Nueva España* by Fray Bernardino de Sahagún (2000), do not mention neither whales, nor dolphins or seals.

The descriptions of the animals of New Spain in the historic chronicles of the 16th century follow the classification of organisms by the ancient Greek philosopher and scientist Aristoteles, but also the emic and vernacular

[1] Landa 1990:157.

[2] An excellent review of historic reports about the Caribbean manatee can be found in McKillop (1985:338-339).

taxonomy used by the Indian population (see Bustos 1988). As a consequence, aquatic mammals were conceived and described as *fishes*, together with other animals that inhabit seas and rivers. Nevertheless, some chroniclers did encounter problems with such a classification, stumbling across the 'cattle-like' properties of manatees, both in their behavior and meat (de Acosta 2003[1589]:179)[3].

An important source that describes manatees in detail was written by the Spaniard Francisco Hernández de Toledo. This chronicler arrived in the 16th century at the Mexican territory under orders of King Phillip II, to realize a study about the natural wealth of this newly acquired land. As a result of his study, Hernández wrote a pivotal opus with the title *Historia Natural de las Cosas de la Nueva España* (Hernández 1959), which includes important information about sea mammals in New Spain. In the third volume, fifth section, chapter LI, he refers to the sea cows as follows:

"Of the SEA COW. In one and another ocean and in the lakes, there lives a fish called by the Haitians 'manatí', it is an almost formless beast similar to a calf, with its head and front arms enlarged like a goat; it is of dun color, is covered with thin hair, and though it is fierce it does not bite. It lives both in the sea and near the beach (and still sometimes it moves away from shore), and feeds of the coastal grasses and of a certain kind of marine fig. It has a rounded tail placed transversely, the head and snout are like a calf, with big nostrils, breasts, ears and small eyes, small teeth also, rough lips, and the skin thicker and stronger than that of a bull. It has two arms at its front part, each one with five nails similar to the humans (...). The navel and the anus are wide, the vulva is like that of woman and the penis as that of a horse. Its fat and meat are like pork, and it is of agreeable taste, whether eaten fresh or salty, harmful (...) for those who suffer from gallstone. The ribs and the entrails are like of a bull but monstrous in size. (...) They procreate a single fetus, since it is very big (...). In the head of this animal there is a little rock, which, triturated and mixed with water or another appetitive drink, expels the retained urine and still everything else that obstructs its conduits. The little rock that is situated in the macho is white and to be used by the men; that of the female is slightly yellowish and useful to the women" (translation by B. Rodríguez).

[3] En las islas de Barlovento, que son de Cuba, la Española, Puerto Rico, Jamaica, se halla el que llaman manatí, extraño género de pescado, si pescado se puede llamar animal que para vivos sus hijos, y tiene tetas y leche con que los cría, y pace yerba en el campo, pero en efecto habita en ordinario en el agua, y por eso le comen por pescado, aunque yo cuando en Santo Domingo lo comí un viernes, cuasi tenía escrúpulos, no tanto por lo dicho, como porque el color y sabor no parecían sino tajadas de ternera, y en parte de pernil, las postas de este pescado; es grande como una vaca (José de Acosta:179)

As one of the first published works composed about the new lands discovered by Cristóbal Colón for the Spanish crown, Gonzalo Fernández de Oviedo (1851[1535]:424-437) wrote in his '*Historia general y natural de las Indias, islas y tierra firme del mar océano*' also about sea wolves (seals), whales and manatees that had been seen in the Caribbean waters. The whales, he stated, were not smaller than the ones hunted at Spain's coasts (ibid.:425), and while Fernández de Oviedo exposes no evidence that those animals were hunted by the Indian fishermen, he does describe an encounter with an apparently female whale and its young in the Gulf of Panama. Seals were, according to Fernández de Oviedo (ibid.:428), frequent inhabitants of the island's and coastal waters. The Spaniards often caught seals while these slept (and snored!) at the shore, using their hides, meat and fat, the latter for lamps in Spain and New Spain. Although Fernández de Oviedo does not detail where the seals were most frequent, he refers to a shipwreck at the 'Alacranes' island, a reef 80 km. north of Yucatán, where several castaways witnessed seals fighting with sharks in the water.

While both, whales and seals were already known to Fernández de Oviedo from Spain, he introduces an entirely new animal by describing the manatee, as he stated, "*one of the most noteworthy fish not known before* [translation by C. Götz]" (ibid.:433). After describing the form and size of these animals, and explaining that some animals can grow as large as fifteen feet long and eight palms thick (ibid.:434), Fernández de Oviedo proceeds to detail the hunting method followed by the Taíno fishermen at Hispaniola. According to his account, this method consisted in shooting arrows tied to ropes into their flesh, pulling the animals to the shore after they died or became too weak to fight, as a cause of the extensive blood loss. On the other hand, the author also explains that Indian fishermen used to catch this animal, or other big aquatic animals such as sea turtles, with a method employing a so called "*pexe reverso*", apparently a remora (Echeneidae), that was 'domesticated' and then used to attach to the prey, allowing it to be pulled close to the canoe or shore to be killed (ibid.:435). The chronicler explains very clearly, in a similar manner as José de Acosta's account written some fifty years later (see above), that the meat, "*if one wouldn't have seen it or wouldn't know it better*", could be easily mistaken with beef or lamb, as both its taste and properties are much different to fish (ibid.:434). The hide, meat and fat of manatees were used for different purposes, and Oviedo prizes especially the fat of these animals to fry eggs, given that it never acquires a rancid taste (ibid.:435).

The Indian hunting techniques have been similar in most areas where aquatic mammals lived; spears or arrows were used to hurt those large creatures, hindering them from diving and causing them to bleed to death, such as Fernández de Oviedo describes for the Antilles. José de Acosta, a Jesuit priest active in the evangelization of Peru during the second half of the sixteenth century, wrote in the year 1589 in his *'Historia Natural y Moral de las Indias*' about the hunting process the Indian populations from Florida used to hunt whales, a method that consisted in introducing a spear or pointed stick into the blowholes of those animals (de Acosta 2003[1589]:181)[4].

A similar description is given by Fray Diego de Landa in the *'Relación de las cosas de Yucatán'* (1990[1566]:157; see also Tozzer 1941:191), the only historic source from proper Mesoamerica that mentions an aquatic mammal. Landa (op.cit) reports that manatees, the only aquatic mammals treated, were shot in shallow estuary waters with harpoons that had buoys attached to them, so that the bleeding animal could be localized quickly when died of blood loss[5].

---

[4] "Pero más maravillosa es la pelea que los indios tienen con las ballenas, que cierto es una grandez del Hacedor de todo, dar gente tan flaca como indios habilidad y osadía para tomarse con la más fiera y disforme bestia de cuantas hay en el universo; y no sólo pelear, sino vencer y triunfar tan gallardamente. (....) El estilo que tienen (según me refirieron personas expertas) los indios de la Florida, donde hay gran cantidad de ballenas, es meterse en una canoa o barquilla, que es como artesa, y bogando, llegarse al costado de la ballena, y con gran ligereza salta y sube sobre su cerviz, y allí caballero, aguardando tiempo, mete un palo agudo y recio que trae consigo por la ventana de la nariz de la ballena (...) luego le golpea con otro palo muy bien y le hace entrar bien profundo" (de Acosta 2003[1589]:181).

[5] "Hay muchos manatís en la costa entre Campeche y la Desconocida, de los cuales, allende del mucho pescado o carne que tienen, hacen mucha manteca y excelente para guisar de comer (...). Lo que yo de ellos digo [es] que son tan grandes que se saca de ellos mucha más carne que de un buen becerro grande, y mucha manteca; engendran como los animales y tienen para ello sus miembros como hombre y mujer, y la hembra pare siempre y no más ni menos, y no pone huevos como los otros pescados; tienen dos alas como brazos fuertes con que nadan, el rostro tiene harta semejanza al buey sácanle fuera del agua al pacer yerba en las orillas, y los suelen picar los murciélagos en una jeta redonda y llana que tienen, que les da vuelta al rostro, y muere de ello porque son muy sanguíneos a maravilla y de cualquier herida se desangran con el agua. La carne es buena, especialmente fresca; con mostaza, es casi como buena vaca. Mátanlos los indios con harpones de esta manera: búscanlos en los esteros y partes bajas, que no es pescado que sabe andar en hondo, y llevan sus harpones atados en sus sogas con boyas al cabo; hallados, los harponean y suéltanles las sogas y las boyas y ellos con el dolor de las heridas huyen a una y otra parte por lo bajo y de poca agua, que jamás van a lo hondo de la mar ni saben, y como son tan grandes van turbando el cieno y tan sanguíneos vánse desangrando; y así con la señal del cieno los siguen en sus barquillas los indios y después los hallan con sus boyas y sacan. Es pesca de mucha recreación y provecho, porque son todos carne y manteca" (Landa 2003[1566]:202-203).

The Archaeological and Paleontological Record of Aquatic Mammals in Southeastern Mesoamerica

As mentioned before, the paleontological and archaeological record of remains from aquatic mammals is relatively scarce in central and southeastern Mesoamerica. Among the numerous pre-Colombian sites of the center of Mexico, the southeastern Gulf of Mexico and the northwestern Caribbean Sea, we could only locate sixteen contexts where fossils or subfossils of aquatic mammal remains were found; many other contexts and sites undoubtedly exist, but might not have been published hitherto. Yet other sites with marine mammal remains, farther south in Central America, can be consulted in McKillop (1985).

The following presentation lists the sites where paleontological or archaeological material of aquatic mammals has been found (Figure 4; see also Figure 5 for a summary of the findings). It includes both coastal and inland sites where ecofacts and artifacts appeared that either derive directly from aquatic mammals or represent aquatic mammals. We define 'ecofacts and artifacts that originate directly from aquatic mammals' as unmodified or indirectly modified archaeofaunal fragments from aquatic mammals, or artifacts carved out of aquatic mammal bone. On the other hand, we include artifacts of any other material that could represent aquatic mammals in our listing, such as paintings or clay figurines. The mentions are ordered by location. The majority of examples comes from coastal sites, simply because the abundance of skeletal material from aquatic mammals has been considerably higher there than at inland sites. True inland sites with archaeofaunal material from aquatic mammals are extremely rare.

Paleontological Finds

The oldest finds of aquatic mammal remains in southeastern Mexico correspond to dugongs (sirenids, comprehending manatees and similar species) and were made at two sites in the present State of Yucatan near the modern coastline, *Colonia Yucatán* and *Noc Ac*. Both paleontological find spots belong to the geological formation called *Carrillo Puerto* (dated between the Late Miocene and the Early Pliocene), approximately between 7 and 5 million years ago (Barnes 2002), and are extremely important for the Mexican paleontological history. Especially the context from Colonia Yucatan is one of the densest and richest finds of sirenid remains in the entire country (Barnes 2002:140), being the dugong remains the only mammal remains of those contexts (Barnes 2002:173). The animals of the taxonomic order of the Sirenia (sirenids) comprise a group of nearly extinct aquatic mammals, with only two remaining genus: the manatee (*Trichechus* spp) that inhabits, among others, the shallow waters of the Gulf of Mexico and Caribbean Sea, and the Dugong (*Dugong* spp), which can be found nowadays in the Pacific, Indian Ocean and Red Sea (Hulbert 2001:322). Sirenids developed during the Eocene from primitive terrestrial animals whose appearance might have been similar to modern proboscides, and that were dwelling in both tropical and artic seas (Hulbert, op.cit). The dugungs persisted living until the Early Pliocene in the coastal waters of the Peninsula of Yucatan, and were later replaced by animals of the Trichechidae family (manatees) during the Pleistocene and Holocene, yet no true fossils[6] of this family have been found in Mexico until now (Barnes 2002:125).

At *Colonia Yucatán,* a nearly complete skull, some vertebras and rib fragments of a young dugong (*Corystosiren varguezi*, a sirenid species similar to the manatee, but with strong and protuberant incisors) were found in a fossil context containing also skeletal remains of several sharks (*Carcharhinus leucas*, *Carcharodon megalodon*, *Isurus hastalis*, and a tiger shark similar to the species *Galeocerdo rosaliensis*) and some bones of prehistoric crocodiles (Barnes 2002; Domning 1990). On the other hand, the *cenote Kambul*, near the actual village of *Noc Ac*, Yucatan, revealed at a depth of approximately 12 meters several fragments of the skull, as well as some loose teeth, of yet another species of dugung (*Xenosiren yucateca*), associated to the remains of prehistoric sharks (*Carcharodon megalodon*, *Hemipristis serra*, *Isurus hastalis*, and *Galeocerdo* sp) (Barnes 2002; Domning 1989).

[6] The archaeological findings of Trichechidae described in the following pages are subfossils rather than true fossils, because the bones still retain their original chemical structure (see Lambert 1979:10-13).

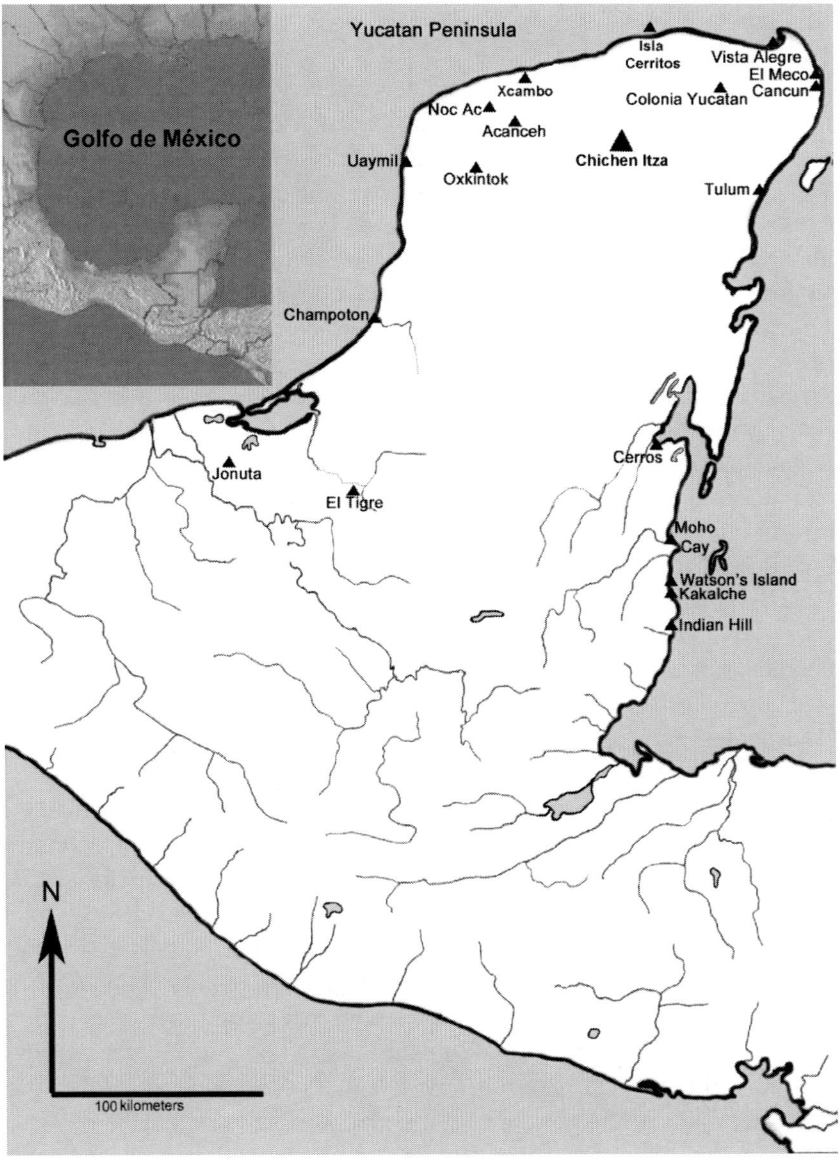

Figure 4. Map of the sites mentioned in this chapter. Drawing by Christopher M. Götz.

Gulf Coast

The zooarchaeological finds from the Maya site of *Champoton* are limited to a sample that was excavated around the Postclassic (AD 1200 until the Spanish conquest) platforms near the actual shore by W.J. Folan and his team of the *Universidad Autónoma de Campeche* between 2003 and 2005. Only four manatee bone fragments were discovered in this sample, three within the contents of a big midden that was formed adjacent to the domestic group no. 5, an important residential group situated relatively near the sea (Götz 2012b), and another adjacent to burial 6 located in the domestic structure 2A of the architectural group no. 8 (Götz 2008a:125). The specimens derive from the appendicular skeleton, the cranium and axial skeleton, none does have cut marks or other traces directly related to human activities, despite their location in archaeological layers.

Another site where manatee remains were found is *Uaymil*, a small island near to Champoton. Uaymil was operated during the Terminal Classic period (AD 850 - 1050) as a transfer port in the pre-Columbian trade system that covered the peninsula's coast (Cobos 2003). At Uaymil, Eaton (1978:57) discovered a manatee rib that was incised with 30 parallel grooves, perpendicular to the longitudinal axis of the rib. The grooves may identify this rib as an *omichicahuaztli*, a prehispanic music instrument similar to the *guiro* used in modern day salsa music of Latin America (see McVicker 2005, and Corona-Martínez, Carmona Macías and Arroyo-Cabrales, this volume).

The third Maya site alongside the Gulf coast where aquatic mammal remains were found is named *El Tigre*. Situated in southern Campeche near the Candelaria River (which discharges into the Gulf of Mexico), this site has an extension of approximately three miles; the topographic map covers 2 miles from East to West and 1.5 miles from North to South (Vargas & Delgado 1999). In El Tigre, about 1700 structures dating from the Late Preclassic (200 BC) to the Late Postclassic (AD 1557) have been registered. The faunal material from El Tigre includes a dolphin tooth (Delphinidae), excavated in the so-called Structure 1. Unfortunately, this tooth was found in a debris area of one of the walls of the structure, so it is not possible to define the reason of its location. It is important to clarify that this marine mammal find relates to other remains of marine fauna found at the site, for example of sea turtles (*Lepidochelis* sp, *Chelonia* sp) and rays (*Dasyatis* sp), which might indicate that the site's inhabitants participated in an exchange network with the southern coast of the Gulf of Mexico.

A fourth Maya site on the Gulf coast where an archaeological specimen of an aquatic mammal was found is *Jonuta*, located in the present Mexican State of Tabasco. The archaeofaunal material, excavated in a test pit as part of the project "The northwestern lowlands of the Maya area" under the direction of Lorenzo Ochoa (Centro de Estudios Mayas of the Universidad Nacional Autónoma de México), was analyzed by biologist Norma Valentín of the Laboratory of Paleozoology of the Instituto Nacional de Antropología e Historia (Mexico). Among the specimens, Valentin (1997) discovered a worn-out rib of sea cow, in the shape of forearm and hand with the palm upwards (Valentin 1997: 348). No further information about aquatic mammal remains was available.

Northern Coast of the Yucatan Peninsula

At the northern coast of Yucatan, there are several sites with active, though also relatively limited evidence of pre-Hispanic exploitation of aquatic mammals. The three sites where archaeological specimens of marine mammals were found did not only reveal numerous remains of manatees, but also of seals, animals that probably were abundant in prehispanic times in the area.

Xcambo, excavated in the 1990's by the Instituto Nacional de Antropología e Historia (Mexico) by Sierra Sosa, is an archaeological site situated on an island in the middle of ancient salty marshes, approximately 700 by 150 meters in east-west direction (Sierra Sosa 1999:42). The site was inhabited between AD 250 and 700, first as a small fishing village, and later as a long distance salt trading port (Wanner et al. 2007). Xcambo's archaeological manatee and seal remains come up to the date from a relatively small portion east to the site's center (see Götz & Sierra Sosa 2011), where most probably the wealthy inhabitants of the island lived, especially during the Late Classic period (AD 550 - 700/850). Despite its proximity to the sea, only two vertebrae fragments and a rib fragment from manatee were found in the platform fills of Xcambo's western part (Götz 2008a:175; Götz & Sierra Sosa 2011:129). Recently, Jiménez Cano (2009) discovered a nearly complete manatee humerus that was chewed by a dog at its proximal end and damaged by surface fire on its medial surface (Figure 5). To this list adds a small number of artifacts, presumably carved out of manatee bone, identified by Canto Méndez (2009: catalogue number 117 (incised spindle whorl), 152 (197) (incised spindle whorl), 174 (polished distal end of a rib), 215 (polished spindle whorl), 221 (polished spindle whorl[7]), and 243 (centrally perforated vertebrae epiphysis)). Among the artifacts, Canto Méndez (2009: catalogue number 262) found also a vertebrae epiphyseal disc presumably from a medium sized dolphin (Delphinidae) that received two perforations on one side as if it was used as a pendant. Furthermore, a vertebra of an unidentified whale was found in the construction fill of the west plaza (Figure 6).

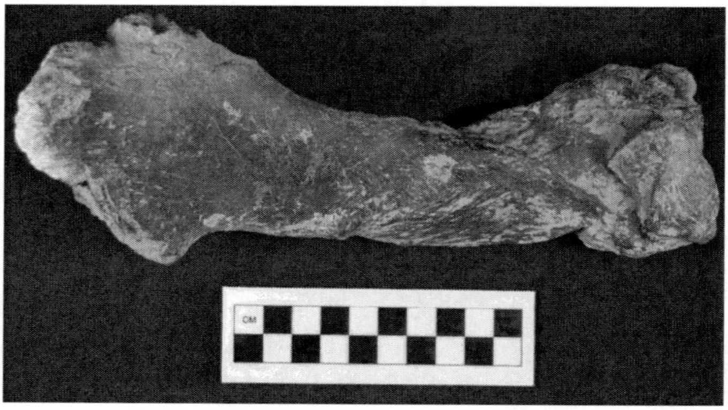

Figure 5. A manatee humerus found in Xcambo. Photography by Christopher M. Götz.

On the other hand, seal remains were relatively abundant at the site; 9 skeletal specimens from the West Indian Seal (*Monachus tropicalis*) were discovered at Xcambo (Götz 2011; Götz & Sierra Sosa 2011). The seal specimens from Xcambo correspond to fragments of the appendicular, axial and cranial skeleton, including a complete femur (Figure 7) and a complete humerus, two radius, a metapodius, as well as a maxilla and zygomatic fragment, together with two loose teeth (see Götz 2012a for a detailed description of the taphonomy and morphometry of the specimens). Although none of the fragments bore direct traces of human activities, all were found in the construction fill of the domestic platforms west of the site's center (as were the manatee bones), and some specimens showed tooth marks attributed to dogs. Additionally to the unworked seal specimens found in the construction fills, Canto Méndez (2009) reports three seal teeth that were perforated or otherwise worked, probably to be used as ornaments. These teeth were found associated to structures NO2, NE3, and NE7A (Canto Méndez 2009: Appendix p. 7, table 'Pendiente')

[7] This artifact was identified by Canto Méndez (2009) as Cheloniidae, but the physical characteristics seem to identify it rather as a spindle whorl carved out of manatee bone.

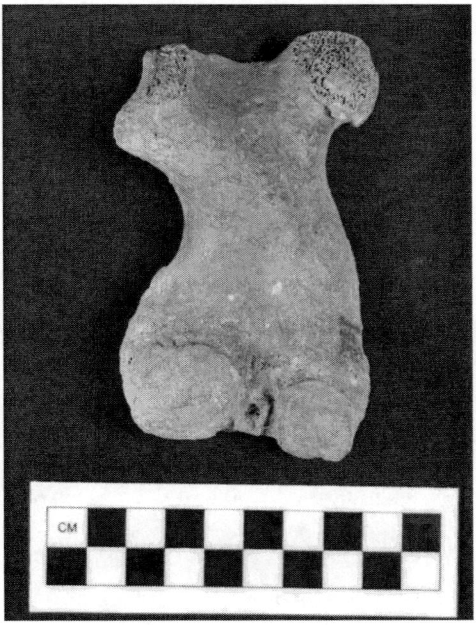

Figure 7. A West Indian seal femur found in Xcambo. Photography by Christopher M. Götz.

The situation for *Isla Cerritos,* a small island occupied approximately between AD 900 and 1,050 apparently by traders with a close connection to the nearby urban center of Chichen Itza (Cobos 2004; Cobos et al. 2007:41, 105; Gallareta Negrón 1998:29), is similar to Xcambo, although less faunal specimens were found. At Isla Cerritos, two excavation campaigns were undertaken, one between 1984 and 1985 by the CRY-INAH and the New College of the University of South Florida (now New College of Florida) (Andrews et al., 1988; Gallareta Negrón 1998; Gallareta Negrón et al., 1989), and another between 2006 and 2007 by the Facultad de Ciencias Antropológicas, Universidad Autónoma de Yucatan (Cobos et al., 2007). The zooarchaeological sample from the 1984/1985 excavations was analyzed by H. Sorraya Carr. In her detailed study, Carr (1987) states that she found 69 (+1) fragments of manatee bone in the Isla Cerritos sample. The manatee bones from Carr's sample represent fairly the whole skeleton, since fragments of the skull (n = 1), vertebrae (n = 7), ribs (n = 15), limb girdle (n = 1) and limbs (n = 41) were found among the material (Carr 1987:19, Table 8). Only three vertebrae fragments showed gouges, scrapes or scratches that Carr (1987:27, table 11) identified as butchery marks, but a total of 49 fragments (71% of the total NISP for manatee) were thermoalterated (6% probably burned, 9% medium burned, 43 % blackened and 13% calcinated) (Carr 1987:32). The zooarchaeological sample excavated at Isla Cerritos between 2006 and 2007 revealed a femur fragment from West Indian Seal, together with an appendicular bone fragment and a vertebrae fragment from a manatee (see Götz 2011 and 2012a for a detailed description and photos). Neither of the specimens showed anthropogenic marks, but a dog apparently chewed the humerus fragment.

Figure 6. A whale vertebra found in Xcambo. Photography by Christopher M. Götz.

At the site of *Vista Alegre,* situated on a small peninsula at the east of the northern coast of the Peninsula of Yucatan and occupied during the Terminal Classic period, only two specimens of marine mammal bones were discovered; one manatee rib fragment and one distal fragment of a seal humerus (García & Götz ms.).

Yet not further detailed, Eaton (1978:9) states that he has found numerous manatee bone fragments in the 35 middens that were surveyed during his pivotal study of the northern and western shore of the Yucatan peninsula. Although there is no detailed information about exactly what kind of sea cow bones were found by Eaton, he does report a manatee rib carved into a music instrument, from Uaymil island, Campeche (see above).

Caribbean Coast

The first archaeological site of the Caribbean where remains from sea mammals were found corresponds to a Postclassic settlement named *El Meco*, which consists in a cluster of small structures, nowadays situated at the interior of the modern city of Cancun, Quintana Roo. Rescue excavations carried out during the 70's and a few studies undertaken in 1980 revealed some isolated remains of a sea cow, associated to a burial. Although it seemed certain that these remains were associated to the funeral context, it was not possible to retrieve any further information, due to the nature of the rescue excavation. The finds were dated to the late Postclassic. Nearby, Wing (1975:186) found one manatee vertebra at a Late Preclassic (450 BC - AD 150) shell midden situated at present day *Cancun*.

The second site is of which archaeozoological material from aquatic animals exists is *Tulum*, a site situated at approximately 80 miles to the south of Cancun on the shore of the Caribbean Sea. Investigations in the 70's revealed a sea cow innominate (pelvis) fragment from this site. The context information indicates that the innominate was found in the fill material of terrace no. 2, but no additional data is available.

McKillop (1985:340-342) lists also three other sites alongside the Caribbean coast of Belize, where manatee bones were found: *Watson's Island* (1 manatee bone fragment from an Early Classic (AD 300 - 550) midden and 2 fragments from a Classic Period deposit), *Kakalche* (5 manatee bone fragments found throughout the site's strata) and *Indian Hill* (several manatee bone fragments found in the archaeological layers of the settlement).

Among the archaeofaunal material from the Preclassic site of *Cerros*, Belize, only a few skeletal fragments from aquatic mammals could be found. Carr, who analyzed the archaeological faunal remains for her doctoral thesis (1986), identified all the aquatic mammal remains as from manatees (see op.cit.: 235, 265, table 15). The skeletal material from manatee found at the so-called 'dock' and at other locations at the site includes a skull fragment, a rib fragment, a loose tooth, and another unidentified piece of bone (Carr 1986:531, 555, 558, 604). Noteworthy is, apart from the unmodified fragments, a single 'thick and polished bone' with scratches and blackened edges that was found in a debitage deposit on the dock (Carr 1986:622).

An important source of information about the exploitation of manatees by the prehispanic Maya comes from the site of *Moho Cay*, explored and excavated by McKillop (2004) in the 80's. At Moho Cay, a small settlement of ancient Maya traders (see McKillop 2004) situated in the mouth of the Belize River and occupied between the Late Preclassic and the Postclassic Period, manatee bones were found in several areas. The first report of literally "tons of manatee bones" from the site comes from Thomas Gann, who explored Moho Cay in the early XXth century (McKillop 2004:260). The high number of manatee bones at the site led Craig (1966) in a description about the geography of fishing in British Honduras and adjacent coastal areas to the conclusion that Moho Cay was a 'Maya fishing site' of people interested in manatee hunting (McKillop 2004:261).

During McKillop's scientific work at Moho Cay, more manatee bones were discovered, especially in a midden named 'Feature 7', where several specimens appeared together with lithic debitage and ceramic shards, identifying the context and an accumulation of anthropogenic material. The manatee bones of the midden were scattered and broken without an apparent pattern, and no signs of butchering could be identified (McKillop 2004:267). Yet, the site of Moho Cay did not only reveal unworked manatee bones from the apparent midden, but also a number of objects carved out of sea cow bones, such as *omichicahuaztlis* (see above), miniature canoe models, fishing weights and human figurines (McKillop 1985:342-345). Those objects, although not depicting a manatee or aquatic mammal themselves, do indicate the importance, at least of this taxon, for sculptural carving.

Inland Sites

One of the oldest and perhaps most intriguing finds related to aquatic mammals far from the sea comes from the Olmec site of *La Venta* in Tabasco, occupied between 1300 and 200 BC. The iconographic representation of what has been identified either as a "dolphin" or a "cetacean" is now exhibited in the *Parque del Museo La Venta* in Villahermosa (see La Crónica 2013). The artifact was presumably among the findings discovered in 1925 by Franz Blom and Oliver La Farge, and shows a remarkable similarity to an orca (*Orcinus orca*), although it has been identified on occasion as a sea cow.

From the area under study, only two true inland sites with archaeological remains related to aquatic mammals could be located. The first of those sites is the prehispanic city of *Oxkintok*, located at more than 50 km from the western coast of the peninsula. Oxkintok was a middle sized Maya settlement, mainly occupied between the Early and Late Classic Period. A Middle Classic (AD 550 - 650) offering in form of a large ceramic vessel found in one of the main architectural groups of the site, the *May* group, included among multiple bone fragments, traces of copal and several jade and shell beads, also a figure of a female individual, carved out of manatee bone (Varela Torrecilla 1991:113,114,116; Figure 8).

On the other hand, in a cave near the modern village of *Acanceh*, Yucatan, Mexico, a small clay flute in form of a seal was found, dated presumably to the Classic Period, according to the discoverer (El Universal 2012). Unfortunately, neither the data regarding the flute nor referring to another supposed painting in yet another cave that apparently shows a manatee have been published yet, so any further analysis will have to wait until this data becomes available to the general public.

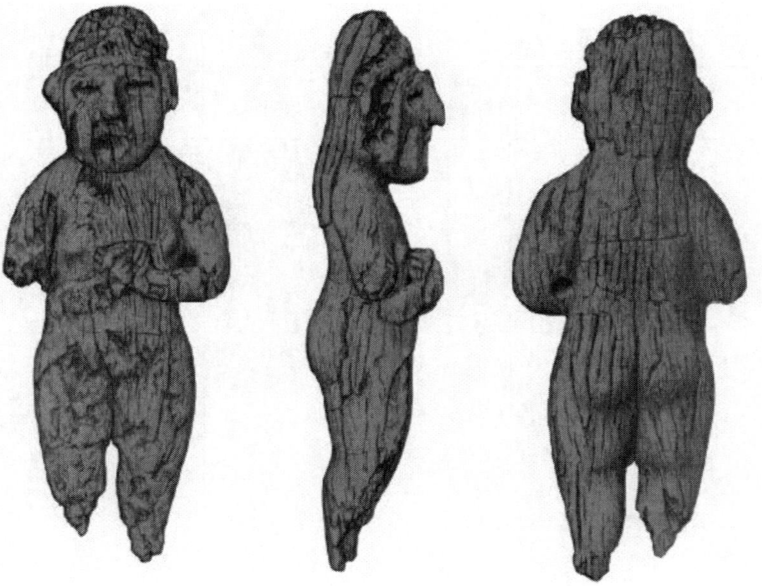

Figure 8. Drawing of the manatee sculpture found in Oxkintok, used with permission of Dr. Miguel Rivera Dorado, director of the Oxkintok Archaeological Project.

One of the few examples of figurines of aquatic mammals found in archaeological contexts in central Mexico consists in a small clay figure discovered in 1999 by Enrique Martínez at the site of *Zultepec-Tecoaque*, Tlaxcala (dating to the 16th century). The figure displays most probably a pinniped or a marine wolf, but unfortunately there is no further information about who had elaborated it, if a Spaniard or an Indian, neither where it had been crafted. The scarcity of data concerning this artifact limits the possibilities of understanding the reason of its presence in the Mexican highlands.

One final example of a possible representation of a sea mammal comes from a Mixtec codex that was elaborated shortly before the Spanish Conquest in Oaxaca, the *Nutall Codex.* On folio 9 (Nuttall 1975:folio 9), there is a scene with a person identified as '4 Ocelotl', who holds in one hand an apparent copal ball, while sustaining with the other an animal which, according to Hairr (2011:21), looks like a seal. Nevertheless, this representation is the only one of its kind in the codex, and no further information can be found to intend to explain the scenery.

Site	Context	Taxon	Specimen(s)	Comment
Colonia Yucatan	Fossil context inside a cenote (sinkhole)	*Corystosiren varguezi*	Nearly complete skull, vertebrae and rib fragments	Oldest finds, not related to human presence (Late Miocene, Early Pliocene)
Noc Ac	Fossil context inside a cenote (sinkhole)	*Xenosiren yucatanica*	Nearly complete skull, loose teeth	Oldest finds, not related to human presence (Late Miocene, Early Pliocene)
Champoton	Domestic midden associated to architectural group 5	*Trichechus manatus*	1 lateral fragment of a lumbar vertebra 1 cranial fragment 1 long bone fragment	The specimens were discovered as part of the midden, which included a high taxonomic variety (see Götz 2012b). None of the specimens showed butchering marks or traces of thermoalteration.
	Structure 2A, architectural group 8	*Trichechus manatus*	1 fragment of a axial vertebra	Although this specimen was found near a human burial (no. 6), it is possible that the specimen only formed part of the soil that covered the context.
Uaymil	Specimen found in Uaymil island.	*Trichechus manatus*	1 rib, almost complete	The specimen has 30 parallel grooves, perpendicular to the long axis; it might have served as a music instrument (see McVicker 2005 and Corona et al. this book).
El Tigre	Structure 1, west wall of building	Delphinidae	Caniniform tooth	The specimen was found isolated in the soil matrix of the debris area.
Jonuta	Test pits 1-4	*Trichechus manatus*	1 rib, almost complete	Worn-out bone carved in form of a forearm with hand, possibly a sort of scepter.
Xcambo	Construction fill of the west part of the settlement	*Trichechus manatus*	2 vertebrae fragments, 1 rib fragment	The specimens were found together with a very diverse collection of faunal material,

Site	Context	Taxon	Specimen(s)	Comment
				none of the manatee specimens had traces of anthropogenic activities.
	Structure NE22, patio XII, square 35-36-39, level I	*Trichechus manatus*	1 nearly complete humerus	The specimen was found together with a very diverse collection of faunal material, it did not have any traces of anthropogenic activities.
	Domestic structures of the east side of the site (Surface, NE2, NE 14A, NE21, NE28)	*Trichechus manatus*	2 incised spindle whorls (cat. no. 117 and cat. no. 152 (197)) 2 polished spindle whorls (cat. no. 215 and cat. no. 221) 1 polished distal end of a manatee rib (cat. no. 174) 1 centrally perforated vertebrae epiphysis (cat. no. 243)	The spindle whorls were found together with a diverse collection of faunal material in the general construction fill matrix of the site.
	Plaza, test pit no. 50, level II	Unidentified cetacean	1 vertebra centrum, spines broken	The specimen was found together with a very diverse collection of faunal material, it did not have any traces of anthropogenic activities.
	Structure NE14	Delphinidae	1 epiphyseal disc of a vertebra with two perforations, possibly to be used as pendant.	The specimen was found together with a very diverse collection of faunal material, it did not have any traces of anthropogenic activities.
	Platform and construction fill of the west side of the settlement (see Götz and Sierra Sosa 2011 and Götz 2012a for details)	*Monachus tropicalis*	1 left femur (ad.), 1 left humerus (ad.), 1 left radius (ad.), 1 right radius (juv.), 2 metapodials, 1 maxillar, 1 canine (associated to the maxilla), 1 loose canine, 1 complete temporal bone with auditorial bulla and zygomatic arch, 3 worked teeth	The specimens were found together with a very diverse collection of faunal material, none of the seal specimens had traces of anthropogenic activities, except for the juvenile radius fragment, which was chewed by a dog.

Figure 9. (Continued)

Site	Context	*Taxon*	Specimen(s)	Comment
Isla Cerritos	Test pits of the excavations of the 80's	*Trichechus manatus*	1 skull fragment, 7 vertebrae fragments, 15 rib fragments, 1 limb girdle fragment, 41 limb fragments	Only a limited number of the specimens showed butchering marks, but nearly 3/4 of the sample was thermoalterated; some specimens were severely burnt.
	Excavations of UADY	*Trichechus manatus*	1 limb fragment, 1 vertebra fragment	Neither of the two specimens had butchering nor fire marks.
	Excavations of UADY	*Monachus tropicalis*	1 femur fragment	A dog or a similar canid chewed the specimen.
Vista Alegre	Test pits distributed throughout the site.	*Trichechus manatus*	1 rib fragment	The specimens did neither show butchering nor fire marks.
	Test pits distributed throughout the site.	*Monachus tropicalis*	1 humerus fragment	The specimens did neither show butchering nor fire marks.
El Meco	Burial 2	*Trichechus manatus*	Fragment of rib, diaphysis of a scapula, and a humerus fragment	Remains associated with a burial near a small structure
Tulum	Terrace 2	*Trichechus manatus*	1 medium part of left pelvis	Isolated element
Cancun	Late Preclassic shell midden	*Trichechus manatus*	1 vertebra	Isolated element
Watson's Island	Early Classic midden	*Trichechus manatus*	3 fragments (unspecified)	No further information.
Kakalche	Throughout the site	*Trichechus manatus*	5 fragments (unspecified)	No further information.
Indian Hill	Throughout the site	*Trichechus manatus*	several fragments (unspecified)	No further information.
Cerros	Excavations associated to the so-called dock and other localities at the site.	*Trichechus manatus*	1 skull fragment, 1 rib fragment, 1 loose tooth, 1 unidentified fragment	According to Carr (1986), some of the specimens were thermoalterated.
Moho Cay	Feature '7'	*Trichechus manatus*	High amount of manatee bones, mainly ribs. At other contexts at the site, several artifacts made of manatee bone were found, such as human figures, canoe models, music instruments and fishing	The breakage pattern of the ribs was identified as natural, no signs of butchering could be found on the bones.

Site	Context	*Taxon*	Specimen(s)	Comment
			weights.	
La Venta	unknown	*Orcinus orca*	Sculpture that resembles an orca (?)	No further information available.
Oxkintok	Group May	*Trichechus manatus*	1 sculpture resembling a naked female figure was found in a ceramic vessel together with copal and jade and shell pendants.	The figurine represents an offering.
Acanceh	Artifact was found inside a cave nearby the modern village.	Focidae	No further information.	The little clay figure is only known from a report in a local newspaper; no image of the specimen is known.
Zultepec-Tecoaque	16th century deposit	Focidae	No further information.	--
Nuttall Codex	Late 15th or early 16th century codex	Focidae	No further information.	Image of a possible seal, sustained by a person identified as '4 Ocelotl'.

Figure 9. Summary of the paleontological and archaeological finds presented in this chapter.

CONCLUSION

In general, there is relatively little information about the role aquatic mammals could have played among the ancient cultures of Mesoamerica. Mesoamerican cultures did actively exploit fauna of the sea and fresh water sources. Coastal settlements of pre-Colombian times clearly indicate that local resources, at coastal lagoons, swamps or other ecosystems near to the land were continuously exploited (Rodríguez 2010), and important numbers of skeletal specimens from both fish and reptiles denote high levels of coastal adaptation (see Götz 2008b, 2012a). Yet, this particular adaptation to coastal environments, so clearly visible in the zooarchaeological record of Mesoamerican sites, shows also that the abundance of fauna in shallow waters was apparently sufficient and that no risk to hunt at the open sea had to be taken (see also Nietschmann (1973), who reported the same pattern for the Miskito of Nicaragua in terms of labor costs hunting manatees vs. collecting molluscs).

In this sense, it is understandable that the two types of sea mammal that displayed the most active signs of exploitation were seals (*Monachus tropicalis*) and manatees (*Trichechus manatus*). Both animals can be hunted in shallow waters or even off the water.

Although manatees are by far the most abundant aquatic mammal in terms of zooarchaeological findings, their remains are scarce in comparison with other frequently used animals in Mesoamerica, such as white-tailed deer, sea turtles or fishes, for example. According to the historical accounts, manatees served as a source for meat, fat, hide and bones. Manatee skeletons are very heavy and solid due to the absence of a medullar cavity in the long bones. The bones of these animals were used for artifacts, and especially the absence of a medullar cavity might have made manatee bones desirable for manufacturers, as many 3 dimensional forms could be sculptured out of such elements. Thus, the scarceness of manatee bones in Mesoamerica's zooarchaeological collections in general could be explained based on those three data: first, manatees were hunted, but many other animal groups could be hunted more easily, second, once hunted, many of the heavy bones were left at the shore or killing site, only transporting the usable bones, fat, meat and hide to the settlements (see Perkins and Daly 1968, but also Baughman 1946), and third, the bones that were actually transported to the settlement, soon entered the exchange market, being sometimes even sent to sites far inland.

Seals, on the other hand, were animals sometimes hunted, but presumably not with such facileness as described by Landa for the manatee. Seals can only be hunted when resting at the beach, as they are vivid swimmers, but stiff and slow on land. Seal populations most probably did not often rest on the beaches of Mesoamerica, but more likely used the many reefs far offshore the Peninsula of Yucatan. Although we cannot know this for sure, modern data from a close relative of the extinct Mesoamerican seal, the *Monachus schauinslandi* from Hawaii, indicate that seals are very sensitive to disturbances caused by people or dogs venturing on the beaches, something that undoubtedly would have been the case throughout the pre-Colombian coast of Mesoamerica in general, and the northern Maya area (where seal bones were found) in particular (Götz 2012a). Yet, at some places in northern Yucatan, seals might have been living near the coast more frequently. Between the modern villages Puerto Progreso and Sisal, in Yucatan (Mexico), there's a small island now in the middle of the salty marshes that constitute the northern part of the peninsula, which bares the Maya name of *Peten Tzula* (Anthony Andrews, personal communication 2011). As *peten* means island, and *tzula* seal, this could mean that at this island, seals regularly rested and perhaps even bred.

On the other hand, the prehispanic fishermen apparently did not consider whales and dolphins as an achievable prey, and the few and isolated remains of those animals found in archaeological contexts correspond most probably to occasional findings of carcasses or bones on the beach. Whales and dolphins, although occurring near the beaches and probably sometimes seen by the people (an indication of such encounters is, for example, the 'orca' sculpture from the Olmec site of la Venta), cannot be hunted easily, and although Mesoamerican fishermen might have had the technique to hunt those animals (such as harpoons, spears, nets and the like), they apparently didn't need to venture into such danger as to hunt an enormous animal from a small canoe, because the 'easily huntable' fauna near the shore was sufficiently abundant to feed the people and meet the needs of fauna for ritual and artifact production.

In this sense, the iconographic data queues into the same line. The rich collections of Mesoamerican wall paintings, sculptures and images on lesser artifacts very rarely depict any taxon that might even have been related to the oceans. In the cases where there are sceneries of aquatic animals (such as known from Chichén Itzá, for example, see Lombardo Ruiz 1998), aquatic *mammals* are commonly *not* among the group of depicted animals. The small seal sculptures found at the post-Conquest site in Tlaxcala and in a cave in Yucatan are rare exceptions, as is the possible drawing of a seal in the Nuttall Codex. If seals, not otherwise known from other that mere coastal sites (Xcambo, Isla Cerritos and Vista Alegre) had a special ritual connotation for the ancient Mayas, has yet to be studied.

In sum, the information presented demonstrates that the Indian people of the central and southeastern part of Mesoamerica somehow knew about the existence of cetaceans, pinnipeds or sirenids, the latter sometimes hunted, the cetaceans only occasionally seen and sculptured, and sometimes found dead on the beach (or parts thereof). All in all, the contacts between aquatic mammals and the Mesoamerican population were only sporadic and no data would confirm such intense exploitation patterns as observed in some south American cultures (for example Zangrando, this volume). One possible explanation could reside in the conformation of the Mexican coasts towards the Gulf of Mexico and Mexican Caribbean, which presented an enormous biological diversity and wealth. From this point of view, the exploitation of the marine fauna in general (mostly smaller animals catchable in nets or from shore) for food or symbolic means existed and is well documented (Götz 2008a; Valadez 1992).

REFERENCES

Adam, P. (2004). Monachus tropicalis. *Mammalian Species 747*, 1-9.

Adam, P., & García, G.G. (2003). New information on the natural history, distribution, and skull size of the extinct (?) West Indian monk seal, *Monachus tropicalis. Marine Mammal Science, 19*(2), 297-317.

Allen, C. (2001). "*Stenella frontalis*". Available at Animal Diversity Web. http://animaldiversity. ummz.umich.edu/accounts/Stenella_frontalis/

Andrews, A. P. (1984). La extinción de la foca (*Monachus tropicalis*) en Yucatán. *Boletín E.C.A.U.D.Y., 12*(68), 3-12.

Andrews, A. P., T. Gallareta Negrón, F. Robles Castellanos, R. Cobos Palma, & Cervera Rivero, P. (1988). Isla Cerritos: an Itzá trading port of the north coast of Yucatán, Mexico. *National Geographic, 4*(2), 196-207.

Barnes, L. (2002). Evolutionary History of the Fossil Marine Mammals of Mexico. In M. Montellano Ballesteros, & J. Arroyo Cabrales (Eds.). *Avances en los estudios paleomastozoológicos* (pp. 125-225). México DF: Instituto Nacional de Antropología e Historia, Colección Científica.

Benton, M. (2000). "*Stenella clymene*". Available at Animal Diversity Web. http://animaldiversity. ummz.umich.edu/accounts/Stenella_clymene/

Baughman, J. L. (1946). Some Early Notices on American Manatees and the Mode of Their Capture. *Journal of Mammalogy 27*(3), 234-239.

Boyle, K. V. (2005). Late Neolithic Seal Hunting in Southern Brittany: A zooarchaeological study of the site of Er Yoh (Morbidian). In G. G. Monks (Ed.). *The Exploitation and Cultural Importance of Sea Mammals* (pp. 77-94). Oxford, GB: Oxbow Books.

Bustos, G. (1988). *Libro de las Descripciones.* Mexico City, Mexico: Universidad Nacional Autónoma de México, Instituto de Investigaciones Históricas.

Canto Méndez, P. M. (2009). *Análisis tipológico y funcional de los artefactos de hueso, asta y diente de Xcambo, Yucatán.* Unpublished BA thesis, presented at the Facultad de Ciencias Antropológicas, Universidad Autónoma de Yucatán. México.

Carr, H. S. (1986). *Faunal Utilization in a Late Preclassic Maya Community at Cerros, Belize.* Unpublished PhD thesis, presented at the Department of Anthropology of the Graduate School of Tulane University.

Carr, H. S. (1987). Preliminary analysis of nonmolluscan faunal remains from Isla Cerritos, Yucatan. Unpublished report, on file at the Laboratorio de Zooarqueología, Facultad de Ciencias Antropológicas, Universidad Autónoma de Yucatán.

Cobos, R. (2003). Classic Maya Seaports: Uaymil, North Campeche Coast. Report submitted to FAMSI. Available at http://www.famsi.org/reports /00013/ 00013Cobos01.pdf

Cobos, R. (2004). Chichén Itzá: Settlement and Hegemony during the Terminal Classic Period. In A.A. Demarest, P. M. Rice, & D. S. Rice (Eds.). *The Terminal Classic in the Maya Lowlands – Collapse, Transition, and Transformation* (pp. 517-544). Boulder, USA: University Press of Colorado.

Cobos, R., Fernández Souza, L., Canto Carrillo, R., Tiesler Blos, V., Cucina, A., Jiménez Alvarez, S., Götz, C., de Anda Alanís, G. & Peniche May, N. (2007). Proyecto arqueológico Isla Cerritos: Estudio de una comunidad maya costera del Clásico Terminal. Unpublished report of the 2006 field season, submitted to the Consejo Nacional de Arqueología, Instituto Nacional de Antropología e Historia. México.

Craig, A. K. (1966). Geography of fishing in British Honduras and adjacent coastal areas. Technical Report on file at the Louisiana State University, Baton Rouge Coastal Studies Institute (Accession Number: AD0636434), 153 pages.

Crónica (Newspaper). (2013). Termina restauración de las 23 esculturas olmecas en el Parque Museo La Venta. Available at http://www.cronica.com.mx/especial.php?id_tema=1226&id_nota=414740.

Davies, L. (2008). "*Monachus tropicalis*". Available at Animal Diversity Web. http://animaldiversity.ummz.umich.edu/site/accounts/information/Monachus_tropicalis.html.

de Acosta, J. (2003). *Historia Natural y Moral de las Indias. Crónicas de América*. Madrid, Spain: Datsin Editores.

Domning, D. P. (1989). Fossil sirenia of the West Atlantic and Caribbean Region. III. *Xenosiren yucateca*, gen. et sp. nov. *Journal of Vertebrate Paleontology*, *9*(4), 429-437.

Domning, D. (1990). Fossil Sirenia of the West Atlantic and Caribbean Region. IV. *Corystosiren varguezi*, gen. et sp. nov. *Journal of Vertebrate Paleontology*, *10*(3), 361-371.

Fernández de Oviedo, G. (1851). Historia general y natural de las Indias. Islas y tierra- firme del mar océano. Available at http://openlibrary.org/books/ OL7096345M/Historia_general_y_natural_ de_las_Indias

Eaton, J. (1978). Archaeological survey of the Yucatan-Campeche coast. *Studies in the Archaeology of Coastal Yucatan and Campeche, Mexico* (1-67). New Orleans, USA: Middle American Research Institute (Publ. 46), Tulane University.

Gallareta Negrón, T. (1998). Isla Cerritos, Yucatán. *Arqueología Mexicana, VI* (33), 24-31.

Gallareta Negrón, T., Andrews, A. P., Robles Castellanos, F., Cobos Palma, R. & Cervera Rivero, P. (1989). Isla Cerritos: Un puerto maya prehispánico en la costa norte de Yucatán, México. *Memorias del II Coloquio Internacional de Mayistas* (Campeche, 1987), *I*, 311-332.

García Paz, C., & Götz, C. M. (ms). Informe de Evaluación de la Muestra Faunística del Sitio Arqueológico de Vista Alegre, Quintana Roo, Temporada 2012. Unpublished report on file at the Laboratorio de Zooarqueología, Facultad de Ciencias Antropológicas, Universidad Autónoma de Yucatán and submitted to the Consejo de Arqueología, INAH (México, 2013).

Götz, C. M. (2008a). *Die Verwendung von Wirbeltieren durch die Maya des nördlichen Tieflandes während der Klassik und Postklassik (600 - 1500 n.Chr.).* Rahden, Germany: Verlag Marie Leidorf.

Götz, C. M. (2008b). Coastal and inland patterns of fauna exploitation in the northern Maya lowlands. *Quaternary international, 191*, 154-169.

Götz, C. M. (2011). Archaeological evidence of the West Indian Seal (*Monachus tropicalis*) from the northern coast of the Yucatán Peninsula. In J. Hairr (Ed.), *Caribbean Monk Seals - Lost Seals of the Gulf of Mexico and Caribbean sea* (pp. 158-165). Landisville, USA: Coachwhip Publications.

Götz, C. M. (2012a). Caza y pesca prehispánicas en la costa norte peninsular yucateca. *Ancient Mesoamerica, 23*, 421-439.

Götz, C. M. (2012b). La fauna vertebrada arqueológica de la costa campechana: el caso de Champotón y Uaymil. In R. Cobos Palma (Ed.). *Arqueología de la Costa de Campeche - la época prehispánica* (pp. 97-123). Mérida: Universidad Autónoma de Yucatán.

Götz, C. M., & Sierra Sosa. T. N. (2011). La arqueofauna de Xcambo, Yucatán, México. *Antípoda, Revista de Antropología y Arqueología,13*, 119-145

Hairr, J. (2011). *Caribbean Monk Seals - Lost seals of the Gulf of Mexico and Caribbean Sea.* Landisville, USA: Coachwhip Publications.

Hernández de Toledo, F. (1959). *Obras completas.* Tomo III, Historia Natural de las Cosas de Nueva España. Mexico City, Mexico: Tratado Quinto, Universidad Nacional Autónoma de México.

Hulbert, R. C. Jr. (Ed.). (2001). *The Fossil Vertebrates of Florida.* Gainesville, USA: University Press of Florida.

Jiménez Cano, N. G. (2009). *Ecología y Subsistencia en la Costa Peninsular Yucateca (250–1150 d.C.), un Acercamiento Paleoecológico a través de la ArqueofaunaCcostera*. Unpublished BA thesis, presented at the Facultad de Ciencias Antropológicas, Universidad Autónoma de Yucatan, Mérida, México.

Lambert, M. (1979). *Fossilien*. München, Germany: Delphin Verlag.

de Landa, Fray D. (1990). *Bericht aus Yucatán*. Leipzig, Germany: Reclam Verlag.

Lombardo de Ruiz, S. (1998). La navegación en la iconografía maya. *Arqueología Mexicana*, *33*, 40-47.

Mangel, J. C. (2003). *Spatial and temporal haul out patterns of the Hawaiian Monk Seal (Monachus schauinslandi) at Midway Atoll, 1997 - 2000*. Unpublished MA thesis, presented at the Nicholas School of Environmental and Earth Sciences de la Duke University, USA.

McKillop, H. (1985). Prehistoric Exploitation of the Manatee in the Maya and Circum-Caribbean Areas. *World Archaeology*, *16*(3), 337-353.

McKillop, H. (2004). The Classic maya Trading Port of Moho Cay. In J.F. Garber (Ed.). *The Ancient Maya of the Belize Valley - Half a Century of Archaeological Research* (pp. 257-272). Gainesville, USA: University Press of Florida.

McVicker, D. (2005). Notched human bones from Mesoamerica. *Mesoamerican Voices, 2*, 1-31.

Nietschman, B. (1973). *Between Land and Water: the Subsistence Ecology of Miskito Indians, Eastern Nicaragua*. New York, USA: Seminar Press.

Nuttall, Z. (1975). *The Codex Nuttall - A picture manuscript from Ancient Mexico*. New York, USA: Dover Publications.

Reid, F. A. (1997). *A Field Guide to the Mammals of Central America and Southeastern Mexico*. New York, USA: Oxford University Press.

Rodríguez Galicia, B. (2010). *Captura, preparación y uso diferencial de la ictiofauna encontrada en el sitio arqueológico de Teopancazco, Teotihuacan*. Unpublished PhD thesis, presented at the Facultad de Filosofía y Letras, Universidad Nacional Autónoma de México.

Sabin, R. (2005). From the paleolithic to the present-day: The Research value of marine mammal remains from archaeological contexts and the uses of contemporary museum reference collections. In G. G. Monks (Ed.). *The Exploitation and Cultural Importance of Sea Mammals* (pp. 1-5). Oxford, UK: Oxbow Books.

de Sahagún, Fray B. (2000). *Historia general de las cosas de la Nueva España*. México DF: Cien de México.

Scott, R. (2002). "*Eubalaena glacialis*". Animal Diversity Web. Available at http://animaldiversity.ummz. umich.edu/accounts/Eubalaena_glacialis/

Sierra Sosa, T. N. (1999). Xcambo. Codiciado enclave económico del Clásico maya. *Arqueología Mexicana*, *VII* (37), 40-47.

Spalding, M. D. (2004). *A Guide to the Coral Reefs of the Caribbean*. Berkeley, USA: University of California Press.

Stora, J. (2002). Neolithic seal exploitation on the Aland Islands in the Baltic Sea on the Basis of Epiphyseal Fusion Data and Metric Studies. *International Journal of Osteoarchaeology*, *12*, 49-64

Stora, J., & Lougras, L. (2005). Human Exploitation and History of Seals in the Baltic during the Late Holocene. In G. G. Monks (Ed.). *The exploitation and cultural importance of sea mammals* (pp. 95-106). Oxford, UK: Oxbow Books.

Tezozomoc, H.de A. (1987). *Crónica Mexicana*. Mexico City, Mexico: Porrúa.

Tozzer, Alfred M. (Ed. and translator). (1941). *Landa's Relacion de las Cosas de Yucatan: a translation*. Cambridge, USA: Papers of the Peabody Museum of Archaeology and Ethnology (14). Harvard University.

Universal (Newspaper). (2012). Hallan posible centro cultural maya. Available at www.eluniversal. com.mx/notas/843908.html

Valadez Azúa, R. (1992). *Impacto del recurso faunístico en la sociedad teotihuacana*. Unpublished PhD Thesis, presented at the Facultad de Ciencias, Universidad Nacional Autónoma de México.

Valentín, N. (1997). Análisis del material zoológico procedente de los sitios arqueológicos de Jonuta, Tabasco y Campeche, México. In Arroyo Cabrales J. & O. Polaco (Coords.), *Homenaje al profesor Ticul Álvarez* (pp. 343-366). Mexico City, Mexico: Colección Científica, INAH.

Vargas, E. & Delgado, A. (1999). "Descubrimientos recientes en El Tigre". *Los investigadores de la cultura maya,1*(7), 112-133.

Varela Torrecilla, C. (1991). La cerámica de Oxkintok: El misterio de la vida y de la muerte. In *Oxkintok - una ciudad maya de Yucatán. Misión Arqueológica de España en México; Proyecto Oxkintok* (pp. 107-141). Madrid, Spain: Ministerio de Cultura, Dirección General de Bellas Artes y Archivos. Instituto de Conservación y Restauración de Bienes Culturales. Comisión Nacional Quinto Centenario del Descubrimiento de América - Encuentro de Dos Mundos.

Wanner, I. S, T. Sierra, K. Alt & Tiesler, V. (2007). Lifestyle, occupation, and whole bone morphology of the Pre-Hispanic Maya coastal population from Xcambo, Yucatan, Mexico. *International Journal of Osteoarchaeology*, *16*, 1–16.

Wing, E. S. (1975). Vertebrate faunal remains. In W. Andrews, M. P. Simmons & Wing, E. S. (Eds.). *Excavation of an early shell midden on Isla Cancun, Quintana Roo, Mexico* (186-188). New Orleans, USA: Middle American Research Institute (Publ. 31), Tulane University.

In: Neotropical and Caribbean Aquatic Mammals ISBN: 978-1-63321-306-7
Editors: Muñoz, Götz and Ramos Roca

Chapter 3

A PREHISPANIC WHALE RIB FROM MONTE ALBÁN, OAXACA: ITS IMPORTANCE IN THE MESOAMERICAN CONTEXT

***Eduardo Corona-Martínez*[1*], *Martha Carmona Macías*[2†] *and Joaquín Arroyo-Cabrales*[3‡]**
[1]Centro INAH Morelos
[2]Subdirección de Arqueología. Museo Nacional de Antropología, INAH
[3]Subdirección de Laboratorios y Apoyo Académico, INAH

ABSTRACT

During explorations undertaken between 1942 and 1944 by the archaeologist Alfonso Caso in the emblematic Zapotec city of Monte Albán, an offering with diverse materials was found, recorded, and collected. Among the materials was a large-size rib that was tentatively identified as whale rib with a number of deep and regular striations. The specimen was recorded and deposited in the collections of the National Museum of Anthropology (MNA, by its Spanish initials). In 1999 this artifact was chosen for permanent exhibit in the Hall of the Oaxacan Cultures at the MNA, but without any further research on the piece, its

* ecoroma09@gmail.com
† arqueologia_mna@yahoo.com.mx
‡ arromatu@hotmail.com

identification was unknown. Some researchers have suggested it this grooved bone functioned as a musical rasp, also known as *omichicahuaztli* in the *nahuatl* or Aztec language.

Curation endeavors currently on-going in the Hall of the Oaxacan Cultures have provided the opportunity to review the artifact using archaeozoological methods, identifying it as a rib from a rorqual whale (*Balaenoptera* sp.). Based on that identification, this paper presents a hypothesis about how it was obtained, as well as evaluating evidence for its possible musical use, all of which contribute to our consideration that this artifact is an exceptional find within the Mesoamerican cultures, as well as the only cetacean remain identified in such a context.

Keywords: Whale, Zapotec, Monte Albán, México, Archaeozoology

INTRODUCTION

Mesoamerican societies of the past are known to have used the abundant available marine faunal resources. Most often, these were allochthonous since those resources were not nearby, so it was likely that they were acquired through commercial or exchange networks, or by intensive regional exploration projects. The recorded marine fauna mainly comprise some important groups, such as: mollusks (Suárez Díez 1989; Velázquez Castro 2000; Suárez Díez & Velázquez Castro 2010; Velázquez Castro et al. 2011); among the vertebrates, there are fish, turtles, birds, and few caimans and pond crocodiles (i.e.: Ocaña & Álvarez 1999), while among mammals, the presence of manatees is noteworthy (i.e.: Mckillop 1985). All of those groups were used for food purposes, as well as for the production of artifacts and ornaments or as models for diverse artistic representations, including symbolic designs.

The archaeofaunal record of cetacean mammals is well known in areas like Baja California (Fujita & Rosales 2000; Fujita 2006; Erlandson et al. 2008) or, through cultural records such as rock art in the Sierras de San Francisco, Guadalupe and La Laguna (Gutiérrez 1994; Pacheco 2006), or in the early historical chronicles (Del Barco 1988; Bernabeu 1994), this knowledge is present also in the coastal cultures from both North American Pacific and Atlantic, where there have been evidences of cetaceans use, either as symbolic elements in sculptures or as utilitarian objects made out of bones: harpoons, scrapers, and hooks, among others (Betts 2007; Erlandson et al. 2008; and references therein).

However, in the well-known and important Mesoamerican cultures, such as the Olmecs, Teotihuacans, Mayas, and Mexicas, there are no evidence of any relationship with cetaceans either as bone, or depictions in lithic material and other artistic features, or even in the main historical chronicles (e.g. Fray Bernardino de Sahagún 1989, Francisco Hernández 1959). In the case of the Zapotec region, no ethnobiological evidence (Hunn 1998; Marcus & Flannery, 2001) suggests some knowledge or use of cetaceans. Nonetheless, this general lack of evidence for cetacean use by ancient Mesoamericans highlights the outstanding nature of the whale rib found by Alfonso Caso in Monte Albán, the capital city of the Zapotec culture (Caso 2006).

ARTIFACT CONTEXT

The archaeological city of Monte Albán has been widely explored since the XIX Century. In 1801 Guillermo Dupaix did an inventory of native Indian antiquities; prior explorations by Henry Holmes in 1884 and by Leopoldo Batres, between 1901 and 1902, added to the site knowledge. However, it was Antonio Caso who did an intensive project of 18 field seasons, exploring and researching, starting with works in 1928 studying the Zapotec hieroglyphs to the extensive excavations he undertook in the 1930s and 1940s, which established the importance of the Zapotec culture as a unique entity with its own characteristics and interaction with contemporary cultures (Winter 2001). Among the explorations done between 1942 and 1944, Caso explored the Northern Platform, excavating stratigraphically a *patio* (Caso 2006), naming it as "*Patio al sur del Montículo A*", with its acronym as *P.S.A.* (Figure 1).

In this patio, Caso (2006) dug 18 test pits. Offering 1 in test pit 3 was found at a depth of 1.52 meters, and contained two whale ribs and a "large number" of flat-based bowls (*cajetes*), several of those incomplete; two bird-shaped pots, one *Quetzalcóatl*–type vessel; one gray-capped vessel; one handle pot; one small bowl; one *apaxtle*; and 23 large rock balls of 14-17 cm diameter. These were generally located over bowls and were surrounded by charcoal remains, but were also found over the *urna* and vessel. There was also a fragment of burned bone, probably vertebrae, inside a small slab stone box. The whale ribs were burned, as if a fire had been set around the offering, and the most complete rib had 70 transversal grooves divided into four groups of 10, 22, 17, and 21 marks (Caso 2006, pp: 125-127) (Figure 2).

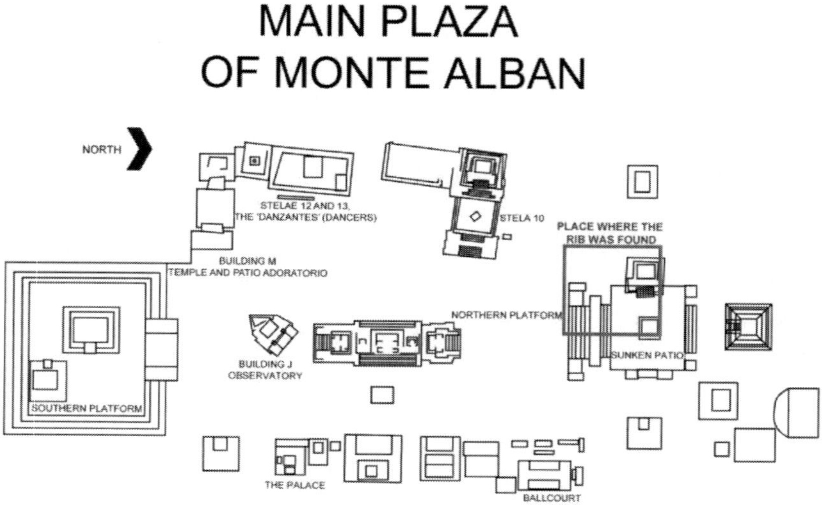

Figure 1. Sketch of Monte Albán; in the box inset is located the finding place. Drawing by Gabriela García Morales (Subdirección de Arqueología, MNA-INAH).

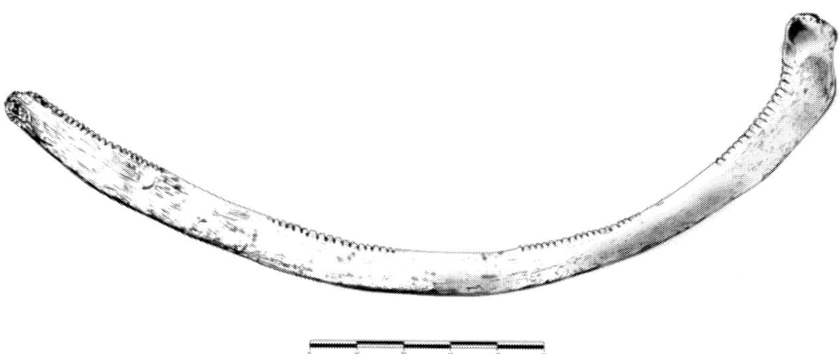

Figure 2. Depiction of the whale rib based on the original work of Caso (2006). Drawing by Alejandra España Paredes, Subdirección de Arqueología, MNA-INAH

This offering was assigned a relative age from the "*época II de la Estructura Azul*" (200 BC - 100 BC), which correlates to the Early Monte Albán II period in the current chronology (100 BC - AD 100), placing it in the urban transition period within the Early Classic (Winter 2001). It is noteworthy to mention that during this period there was contact with Teotihuacan, and an important campaign for expanding the Zapotec state.

The offering materials were deposited in the collections of the National Institute of Anthropology and History (INAH by its initials in Spanish), and in 1964, when the National Museum of Anthropology was inaugurated, the materials were moved to the museum collections. In 1999, with the renovation of the Hall of Oaxacan Cultures, the best preserved whale rib and the one which is studied here, was incorporated to the public exhibition. The other archaeozoological items from the same offering are also under study.

IDENTIFICATION NOTES

Archaeozoological studies have been on-going at INAH since 1963, but these mostly deal with artifacts from prehistoric sites and quaternary paleontological localities (Corona-M. 2008a; Corona-M. et al. 2010). This could explain why the specimen was not previously studied, and instead the preliminary identification provided by Alfonso Caso as a whale rib was retained. That identification was justified due to the huge size of the rib, but there was no further support for the identification based on analytical tools.

More recently we studied the specimen in detail to investigate its possible uses. This study is presented below.

The first stage of the analysis was the morpho-anatomical review of the specimen and its osteometric comparison with modern comparative specimens from several scientific collections. This has been reported elsewhere (Corona-M. et al. 2012a, b), so here we present only a brief synthesis:

The size of the rib was an initial clue that it pertained to a whale, but that is a generalized identification since there is a large diversity of marine mammals, from the enormous blue whale (*Balaenoptera musculus*) and the sperm whale (*Physeter macrocephalus*) through smaller cetaceans, like the pigmy sperm whale (*Kogia* sp.), or the beaked or mesoplodont whales (*Mesoplodon* sp.) with lengths between 2 and 4 m and ribs much smaller the studied one. All of the mentioned cetaceans are currently known from the Mexican Pacific waters (Ceballos & Oliva 2005).

Based on specimen size alone, it was also possible that the artifact could pertain to a large terrestrial mammal like a mammoth (Proboscidea). This option is possible since the XVI Century historical chronicles make it clear that extinct megafauna were known by the Mesoamerican societies, although many, like the Mexicas thought that these remains were those of giants, named *quinametzin* (Corona-M. et al. 2008).

The lack of research at the time of the artifact's recovery with regard to the relationships between ancient Mesoamerican cultures and fossil remains (Corona-M. et al., 2008, Cuevas & Alvarado 2011), is one reason that the original researchers assigned the studied specimen to the whales. When it was recovered in Monte Albán, information about megafauna was concentrated in the Mexican Plateau and the Basin of México. In the last decades, there have been several mammoths recovered in the Mixteca, the Oaxacan region where Monte Albán is located (Pérez-Crespo et al. 2008). Since it is sometimes possible to collect specimens in good condition, it was important to explore the possibility that the specimen could be the remains of those animals, although the following study shows that it is not.

The rib was identified as pertaining to a specimen of a bearded whale or Mysticeti, particularly a fin whale (*Balaenoptera* sp.). Within the Mexican Pacific marine region from southern Baja California peninsula to Central America, this genus has four species: Sei whale (*B. borealis*), Bryde´s whale (*B. edeni*), fin-backed whale (*B. physalus*), and blue whale (*B. musculus*). Ribs do not usually have diagnostic characters to separate these species, and also the taxonomy of the species within the genus is quite complex, in order to made a specific assignation. Thus, we only tentatively assign it to the fin-backed whale (*Balaenoptera* cf. *B. physalus*) (Corona-M. et al. 2012a, b).

The specimen shows some weathering on the cortical portion, but there is no sign of the use of any preservative method post-recovery. In the last century, these methods of preservation were usual, but used to be quite aggressive to the bone. Instead, the specimen is stable and in good condition. As for cultural modifications, there are the four grooves series discussed above (17, 20, 17, 21 counting from the proximal end of the rib; each series with a different depth. There is no other sign of fracture, wearing, or burning. Both the number of grooves and the other characters differ from those mentioned by Caso (2006) (Figure 3).

INTERPRETATION

Using the archaeozoological method, and particularly with a positive identification, there is access to a large amount of biological (such as animal size and natural geographic range) and cultural information (such as collecting and processing procedures) (Polaco 1991; Reitz & Wing 2008). Based on those data, we discuss the rib whale identified.

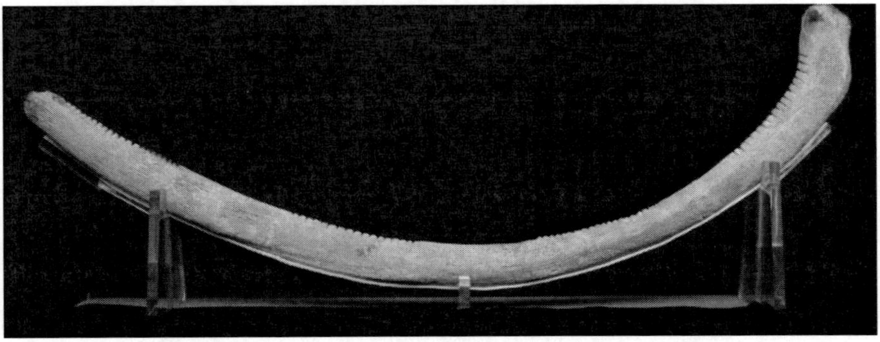

Figure 3. The whale rib of Monte Albán currently exposed in the Hall of Oaxacan Cultures, identified as a fin-backed whale (details in text, photo by Martha Carmona, Subdirección de Arqueología, MNA-INAH).

About the availability of specimen, as mentioned above, fin-backed whales are found in the marine southern Mexican Pacific, a region that is used for migration. In fact, there are some recent records of specimens of the genus *Balaenoptera* on the Oaxacan coasts, such as a visual recording of the Bryde´s whale (*B. edeni*) and a report of a stranded blue whale (*B. musculus*) (Meraz & Sánchez-Díaz 2008). It is important to point out that those cetaceans used to be quite abundant. Documents show that starting the past century, intensive world-wide hunting diminished the population by about 70% between 1929 and 2007, especially in the southern hemisphere. As a result, this species has been assigned a conservation category of "threatened" (Reilly et al. 2013). There is no way to establish a specific availability area since the fin whale distribution in Mexican seas is wide, including both Atlantic and Pacific waters, excepting the Caribbean (Jefferson et al. 1993; SEMARNAT 2003; Heckel et al. 2008). However, we were able to infer a minimum distance for access, using a map and tools provided by Google Maps. The shortest straight-line distance between downtown Monte Albán and the Oaxacan coast is 140 km (Figure 4), suggesting that several days of travel were required to secure the specimen. Additionally, some kind of packing would have been required (Figure 4).

As for the way in which the specimen was obtained, the lack of butchering marks and the weathering suggest a feasible hypothesis that it came from a stranded and decayed carcass that was covered in some portions with skin or sand, providing a differential preservation until its recovery. If that was indeed the case, a specimen recovered by this way would have the required integrity to be transported from the coastal to the main city.

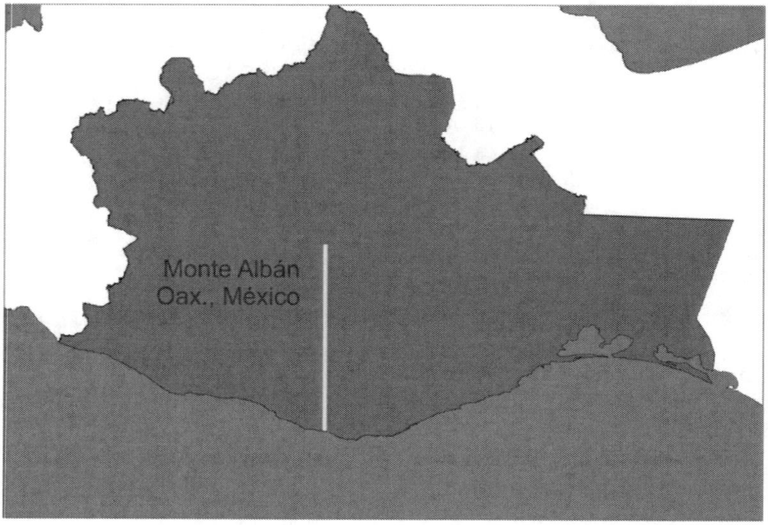

Figure 4. Map of Oaxaca showing the location of Monte Albán (dot) and the minimal distance for access to whale rib in the Pacific coast. The measure of straight line is 140 km, details in text. Figure elaborated by Eduardo Corona-M. and Rafael González.

About the artifact use, it should be remembered that the studied rib is part of an offering context and future interpretation should consider all aspects of the offering context, including any other bone. Some previous studies postulate that this piece is a musical artifact, based on that the grooves are quite similar to musical rasps (Martí 1968; Pareyón 2007), including those particularly named *omichicahuaztli* (Martí 1968). The presence of the artifact in Monte Albán was explained by a vague and uncertain relationship between Yopi people that inhabits in Pacific coast of Guerrero and nearby Oaxacan border, and the Mixteco people, from central Oaxaca (Pareyón 2007).

We consider that term *omichicahuaztli* should be used cautiously since it is a Nahuatl word, and applied to Aztec notched bones played on the occasion of elite mortuary rituals, but in fact were found two different types of rasps in Posclassic of Valley of Mexico, one engraved with images and frequently grooved with deep and regular striations, found in non-mortuary contexts and other notched but not incised and recovered in great numbers from secondary burials (Mcvicker 2005). Then, the use of term *omichicahuaztli* could suggest a false idea of continuity or contemporaneity. The use of musical rasp in several distinct Mesoamerican cultures is extensive, and in every language there could have been a specific term for the implement.

Regardless of the used name, in order to consider the artifact as a musical instrument, its functional use must first be explored. The rib does not show any evident use signal, and in such a case it should include marks caused by the bony or wooden instrument used to scrape it. These should include macroscopic wear marks or, even from an occasional use, some scratching marks. Those marks should certainly be evident because in order to produce sounds, the scraping required a strong effort because this rib does not have a resonance box as in the case of musical rasps made out of long and hollow bones. Even had there have been an external resonance box, and then the use of such a box should have left some marks on the artifacts, and these marks were not found on the rib. For all the reasons cited above, the actual use of the artifact is not clear, and its musical use is questionable.

If the whale rib was not used as a musical instrument, maybe their actual function was something different, such as a ritual or ceremonial instrument or otherwise, such inferences require a careful study of the offering context as well as its relationship with Zapotec cosmovision. An important consideration is the series of patterned marks: 17-20-17-21. We wonder if the actual function of the specimen was as a marker. However, this is just one other possibility among several, and further studies are warranted.

In any case, notwithstanding its actual purpose, the artifact could be classified as a sumptuary item, one with a high value and owned by people of high range in the social hierarchy. Initially in zooarchaeological studies, the sumptuary value of animals was considered initially only in terms of dietary activities (Ervynck et al. 2003). But such sumptuary value is also now well known in Mesoamerican, some animals unavailable on the surrounding habitat or difficult to obtain were used in rituals or in the production of animal-based artifacts (Corona-M. 2008b; Emery 2002). This is certainly the case for the whale rib that was collected at a minimal distance of 140 kms.

CONCLUSION

One unique finding of this study that is not simply an inference is that this archaeological specimen represents the only Cetacean bone recorded in the context of the Mesoamerican cultures. It is thus an exceptional find. As a result, more questions than answers have appeared: Why did these eminent marine mammals not get the interest of Mesoamerican societies? Cetaceans, including dolphins, porpoises and whales, are a common view from the coast, and stranded animals are not that rare in some areas. So, did coastal people not know or get interested in them? Was Prehispanic technology not well enough developed for using such a resource? Why it was not recorded in their cosmovision? Or perhaps, did the Zapotecs not think that it was a whale rib and thought it was something else?

Those questions are particularly useful because the Pacific cultures to the north are well-known to have widely used whales as a resource, so another question is: Has the whale rib arrived from somewhere else? Finally, it was not the purpose of this paper to solve the cultural questions of the whale rib specimen, but to identify the origin of the bone and explore some explanations for its presence at the site. As mentioned before, further studies are warranted, including physico-chemical and dating assays, and also anthropological analyses of the offerings to understand this piece as part of the Zapotec cosmovision.

Acknowledgements

The authors were grateful for the kind invitation from the editors to contribute a manuscript for this volume, and especially to Chris Götz for his patience. To the reviewers, their recommendations helped to improve the original manuscript. And to Kitty F. Emery (Natural History Florida Museum), her support to fight against baroque Spanish sentences and suggest to us a nice English translation of the manuscript, but finally any mistake found is responsibility of the authors.

References

Álvarez, T. & Ocaña, A. (1999). *Sinopsis de restos arqueozoológicos de vertebrados terrestres, Colección Científica 386*. Mexico City, Mexico: Instituto Nacional de Antropología e Historia.

Bernabeu, A. S. (1994). *"Diario de las expediciones a las Californias" de José Longinos*. Madrid, Spain: Editorial Doce Calles.

Betts, M. W. (2007). The Mackenzie Inuit whale bone Industry: Raw material, tool manufacture, scheduling, and trade. *Arctic 60*(2), 129– 144.

Caso, A. (2006). *Obras: El México Antiguo. Volumen 4: Mixtecas y Zapotecas*. Mexico City, Mexico: El Colegio Nacional.

Corona-M. E. (2008a). The origin of archaeozoology in México: An overview. *Quaternary International, 185*(1), 75–81.

Corona-M. E. (2008b). Zoogeographical affinities and the use of vertebrates in Xochicalco (Morelos, Mexico). *Quaternary International, 180*(1), 145-151.

Corona-M. E., J. Arroyo Cabrales & Polaco, O. (2010). La Arqueozoología en México, una reseña actual. In G. Mengoni Goñalons, J. Arroyo Cabrales, Ó. Polaco & F. J. Aguilar (Eds.). *Estado actual de la arqueozoología latinoamericana/ Current advances in Latin-American archaeozoology* (pp. 165-172). Mexico City, Mexico: Instituto Nacional de Antropología e Historia. Consejo Nacional de Ciencia y Tecnología. International Council of Archaeozoology. Universidad de Buenos Aires.

Corona-M. E., J. Arroyo Cabrales & Carmona Macías, M. (2012a). Identificación del hueso trabajado de ballena (Monte Albán, Oaxaca). *El Tlacuache suplemento cultural (Centro INAH Morelos y La Jornada Morelos), 550*, 4.

Corona-M. E., J. Arroyo Cabrales & Carmona Macías, M. (2012b). *Identificación del hueso trabajado de ballena (Monte Albán, Oaxaca).* Report on file at Centro INAH Morelos and Museo Nacional de Antropología, Mexico City, Mexico.

Corona-M. E., M. Montellano Ballesteros & Arroyo Cabrales, J. (2008). A concise history of Mexican paleomammalogy. *Arquivos do Museu Nacional, Rio de Janeiro, 66*(1), 179-189.

Cuevas G., M. & Alvarado, J. (2011). El mar de la creación primordial: un escenario mítico y geológico en Palenque. *Arqueología Mexicana, XIX* (133), 32-37.

Del Barco, M. (1988). *Historia natural y crónica de la antigua California.* Mexico City, Mexico: Universidad Nacional Autónoma de México.

Emery, K. J. (2002). The noble beast: status and differential access to animals in the Maya world. *World Archaeology, 34*(2), 498–515.

Erlandson, J. M., M. L. Moss & Des Lauriers, M. (2008). Life on the edge: early maritime cultures of the Pacific Coast of North America. *Quaternary Science Reviews 27*, 2232–2245.

Ervynck, A., W. Van Neer, H. Hüster-Plogmann & Schibler, J. (2003). Beyond affluence: the zooarchaeology of luxury. *World Archaeology 34*(3), 428–441.

Fujita H. & Rosales López, A. (2000). *La Antigua California prehispánica: La vida costera en El Conchalito*. Mexico City, Mexico: Instituto Nacional de Antropología e Historia.

Fujita, H. (2006). The Cape region In: D. Laylander & Moore, J. D. (Eds.), *The Prehistory of Baja California: Advances in the Archaeology of the Forgotten Peninsula* (pp. 82–98). Gainesville, USA: University Press of Florida.

Gutiérrez Martínez, M. L. (1994). *Arte Rupestre Baja California Sur, la Sierra de San Francisco, el Cañón de Santa Teresa.* Mexico City, Mexico: INAH-SALVAT.

Heckel, G., P. Ladrón De Guevara & Rojas Bracho, L. (2008). Ballenas y delfines. In: G.D. Danemann & E. Ezcurra (Eds.). *Bahía de los Ángeles: recursos naturales y comunidad. Línea base 2007* (pp. 563-600). Mexico City, Mexico: Secretaría de Medio Ambiente y Recursos Naturales. México.

Hernández, F. (1959). *Obras Completas vol. III.* Mexico City, Mexico: Universidad Nacional Autónoma de México.

Hunn, E. S. (1998). Mixtepec-Zapotec Ethnobiological Classification: A Preliminary Sketch and Theoretical Commentary. *Anthropologica, 40*, 35-48.

Jefferson, T.A., S. Leatherwood & Webber, M.A. (1993). *FAO species identification guide. Marine mammals of the world.* Rome, Italy: FAO.

Marcus, J., & Flannery, K. V. (2001). La clasificación de animales y plantas entre los zapotecos del siglo XVI. Un estudio preliminar. *Cuadernos del sur 7*(16), 5- 20.

Martí, S. (1968). *Instrumentos musicales precortesianos, 2ª Ed.*. México City, México: INAH.

Mckillop H. (1985). Prehistoric explotation of the manatee in the Maya and the circum-Caribbean areas. *Journal of World Archaeology, 16*(3), 337-353.

McVicker, D. (2005). Notched human bones from Mesoamerica. *Mesoamerican Voices, 2*, 1-31.

Meraz J., & Sánchez-Díaz, V. M. (2008). Los mamíferos marinos de la costa de Oaxaca. *Revista Mexicana de Biodiversidad 79*, 143-161.

Pacheco Bribiesca, R. C. (2006). *Cultura del Mar: grupos nómadas que habitaron en el pasado la península de Baja California y su relación con las ballenas.* Unpublished BA thesis, presented at the Escuela Nacional de Antropología e Historia, Mexico City, Mexico.

Pareyón, G. (2007). La música en la fiesta del dios XipeTotec. In G. Castillo Ponce (Ed.). *Actas del III Foro Nacional Sobre Música Mexicana: "Música y músicos a lo divino. Homenaje a Miguel Bernal Jiménez".*

Available at https://tuhat.halvi.helsinki.fi/portal/en/publications/la-musica-en-la-fie(d1dc0f87-5199-48d4-8644-a2299f970129).html

Pérez Crespo, V. A., J. Arroyo-Cabrales & Santos Moreno A. (2008). Generalidades de los Mamíferos del Pleistoceno Tardío de Oaxaca. *Naturaleza y Desarrollo 6*(2), 5-11.

Polaco, O. J. (Ed.). (1991). *La Fauna en el Templo Mayor.* Mexico City, Mexico: Asociación de Amigos del Templo Mayor, A. C. Instituto Nacional de Antropología e Historia. García y Valdés Editores.

Reilly, S.B., Bannister, J. L., Best, P. B., Brown, M., Brownell Jr., R. L., Butterworth, D. S., Clapham, P. J., Cooke, J., Donovan, G. P., Urbán, J. & Zerbini A. N. (2013). *Balaenoptera physalus.* Available at http://www.iucnredlist.org/ details/2478/0

Reitz, E. J. & Wing, E. S. (2008). *Zooarchaeology.* Cambridge, UK: Cambridge Manuals in Archaeology.

Sahagún, B. (1989). *Historia General de las cosas de la Nueva España.* 2 vols. Mexico City, Mexico: Consejo Nacional para la Cultura y las Artes. Alianza Editorial Mexicana.

Semarnat. (2003). Informe de la situación del medio ambiente en México, 2002. Cap. 6. Biodiversidad. Available at http://www.paot.org.mx/centro/ inesemarnat/informe02/estadisticas_2000/informe_2000/index.htm

Suárez Díez, L. (1989). *Conchas prehispánicas en México.* Oxford, UK: B.A.R.

Suárez Díez, L., & Velázquez Castro, A. (2010). *Ecos del pasado. Los moluscos arqueológicos de México.* Mexico City, Mexico: Colección Científica 572. Instituto Nacional de Antropología e Historia.

Velázquez Castro, A. (2000). *El simbolismo de los objetos de concha encontrados en las ofrendas del Templo Mayor de Tenochtitlan.* México City, México: Colección Científica 403. Instituto Nacional de Antropología e Historia.

Velázquez Castro, A., Melgar Tísoc, E. R. & Gómez Gastélum, L. (Coords.). (2011). *Moluscos arqueológicos de América.* Guadalajara (Jalisco), Mexico: Universidad de Guadalajara.

Winter, M. (2001). La zona oaxaqueña en el Clásico. In Manzanilla, L. & L. López Luján (Coords.), *Historia Antigua de México, El Horizonte Clásico vol. II.* (pp. 41-64). Mexico City, Mexico: INAH, UNAM, Miguel Ángel Porrúa.

In: Neotropical and Caribbean Aquatic Mammals ISBN: 978-1-63321-306-7
Editors: Muñoz, Götz and Ramos Roca

Chapter 4

Zooarcheological Evidence on the Utilization of Marine Mammals in Northern South America and the Caribbean: A Contribution to Long-Term Biological Conservation

***Elizabeth Ramos Roca*[1,*] *and Lourdes Pérez Iglesias*[2,†]**
[1]Department of Anthropology, Universidad de los Andes, Bogotá, Colombia
[2]Department of Archaeology, Eastern-Central Region, Centro de Investigaciones y Servicios Ambientales y Tecnológicos (CISAT)

Abstract

Among vertebrates, certain large marine mammals possess great potential as a foodstuff for human consumption. However, despite this, and despite their other potential uses, the archaeological evidence concerning their use in northern South America and the Antilles is relatively scarce. This chapter reviews and discusses the available information on the use of marine mammals in this Neotropical region. The area of study has awakened the interest of specialists, particularly in relation to the effects that over-exploitation by humans has had on these

* E-mail: eramosroca@uniandes.edu.co
† E-mail: lourdes@cisat.cu

species, some of which are now extinct, face extinction or are potentially endangered. Concretely, the chapter presents information drawn from archaeological reports from the Caribbean and northern South America and in particular from Colombia, Cuba and the Antilles; it discusses the usefulness of this information for reconstructing the uses marine mammals were put to in the region.

In addition to confirming the scarcity of data on this group of vertebrates the chapter provides evidence of how in some cases the indigenous population considered their capture to be of particular importance, because of the abundant edible biomass they provided and the raw materials extracted from them that were used to manufacture body ornaments and funerary and ceremonial objects.

Keywords: Marine mammals, Zooarchaeology, Colombia, Cuba, the Antilles

INTRODUCTION

The contribution that can be made by archaeological research, and in particular by zooarchaeology, to the reconstruction of topics such as the relation between humans and their environment depends on the nature of the research questions that are posed and the quality (integrity) of the samples recovered and analyzed. Thus, in the particular case of the reconstruction of possible uses of marine mammals, it is first of all necessary to carry out a careful evaluation both of the available evidence (quantity, context, etc.) and of the factors that might have affected the nature of the samples (such as their state of preservation, representativeness, etc.).

The reflections presented in this chapter have their origins in a series of general questions: Were marine mammals an important subsistence resource for the indigenous populations of northern South America and the Antilles? Does sufficient archaeological evidence exist to evaluate the ways in which this resource was used? What is the nature of this evidence? With the available information, is it possible to hazard some generalizations about these matters?

These questions are tied to the fact that in current discussions concerning the preservation of different aquatic mammal species, their indiscriminate use by human populations is offered as one of the fundamental reasons they have come under threat or in some cases become extinct. However, as might be supposed, if these impacts are to be evaluated, a longer term perspective is required, along with adequate data to support them.

Building on these concerns, this chapter seeks to carry out an initial evaluation of the kinds of evidence that are available to evaluate the use made of marine mammals by indigenous communities in northern South America and the Antilles. The first section of the chapter presents the results of the reported data on the use of marine mammals in Cuba and the Antilles that have been examined in the archaeological literature. The second section comprises an archaeological examination of the significance of marine mammals to the aboriginal populations of present-day northern Colombia. Finally, the last section discusses these results and compares them with information available for other regions of the Americas.

The Antilles

The human groups that inhabited the insular area of the Caribbean did not have access to large land animals. The only sizeable terrestrial species reported for the region were members of the *Megalochnidae* (land sloth) family (Order: *Pilosa*), but it is estimated that they were exploited only by the earliest cultures as their extinction has been dated to at least 4,000 years ago (Rodríguez & Vento 1989). A recent analysis carried out by Díaz-Franco (2011) proposes that, despite the possible coexistence in Cuba during the Holocene of aborigines with sloth genera such as *Megalocnus, Mesocnus, Neocnus* and *Acratocnus*, the dating of the available samples cannot be considered reliable as in some cases there is no consensus regarding the origins of the deposits in which the fossils were found, insufficient information is available on the context of the findings and no taphonomic analyses have been carried out on the remains found there.

Other terrestrial vertebrates used by primitive communities in the Antilles included different species of capromys (medium- to small-sized rodents), birds and reptiles, but none of these reached the size of the marine mammals. It might therefore be assumed that marine mammals would have been of great significance for pre-Columbian groups, providing substantial food resources as well as raw materials for the fabrication of tools, body ornaments and ceremonial objects.

Few archaeological studies of marine mammals have been conducted for the Antilles, not because of lack of interest but because of the scarcity of archaeological evidence, which is, furthermore, generally confined to coastal areas. The database, *Arqueofauna de la región Centro Oriental de Cuba*, and the catalog *Fauna de sitios arquelógicos aborígenes de Cuba* were used to compile extensive information on the ancient fauna of Cuba found at archaeological sites, and including a section dedicated to marine mammals (Morales et al. 2011 and Pérez et al. 2012). In addition, the authors carried out a specific bibliographical search on the topic (Figure 1).

Archaeological sites containing evidence of marine mammals

Hitherto, the three species of marine mammals that have been most reported for Cuba are the manatee (*Trichechus manatus*, Linnaeus, 1758), the Caribbean monk seal (*Monachus tropicalis*, Gray, 1850) and the bottlenose dolphin (*Tursiops truncatus*, Montagu, 1821). Rouse (1942) provides the only report of whale remains, reported for the site of Varela III, but it is not clear which species is referred to.

The manatee (*Trichechus manatus*, Linnaeus, 1758), Order *Sirenia*, Family *Trichechidae*, lives in rivers, estuaries and shallow coastal waters. Its diet consists of aquatic plants and it has abundant edible biomass, calculated at approximately 250 kg. It is found from the south eastern United States to the north eastern coast of Brazil. Two sub-species are recognized: the Florida manatee (*Trichechus manatus latirostris*), limited to the south eastern US, and the Antillean manatee (*Trichechus manatus manatus*), which is found in the rest of the Caribbean.

Trichechus manatus remains have been identified at several Cuban archaeological sites such as: Tasajera I (Mayabeque Province), Asiento Carapachibey (Isla de la Juventud), Cayo Jorajuría (Matanzas Province), Cayo Carenas (Cienfuegos Province), Solapa Alta (Villa Clara Province), Toma de Agua, Toma de Agua de Río Zaza, Cayo Lucas, Estero del Macío (Sancti Spiritus Province), Los Buchillones (Ciego de Ávila Province), Victoria I (Camagüey Province), La Victoria (Las Tunas Province), Cayo Bariay, Corinthia III, Esterito, San Antonio, El Mango (Holguín Province), Corrales de Ojo de Toro, Asiento el Carnero (Granma Province), Caimanes III, Caimanes II, Cayo Caimanes, Cocal Largo, Cayo Damas (Santiago de Cuba Province) (Morales et al. 2011) and (Pérez et al. 2012) (Table 1).

Today, the species is absent from many of the locations where its ancient remains have been reported, indicating that the population has declined and/or been displaced to other more protected areas in the face of hunting and habitat degradation. In recent excavations carried out in 1998 and 2008 respectively at the sites of Cayo Bariay, to the west of the Bay of Bariay and San Antonio, east of the Bay of Gibara - both in Holguín Province - abundant remains of this member of the *Sirenia* order were found (Figure 2). It is many years since the species was last observed at either of these sites (Pérez & Guarch, 2003 and Jardines et al. 2009). This situation is not confined to Cuba, but occurs in other Caribbean islands and in Central America too, leading the International Union for Conservation of Nature (IUCN) to categorize the Caribbean manatee as vulnerable, though in Cuba it has been classed by experts as threatened or endangered (Álvarez 2012).

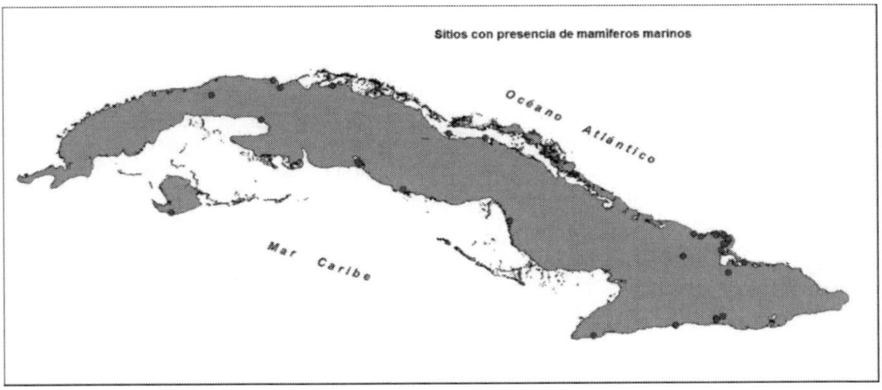

Figure 1. Localization map showing Cuban sites where aquatic mammal remains have been found.

Figure 2. Manatee rib found at a depth of 70 cm at the San Antonio site (to the east of the Bay of Gibara, Holguín, Cuba). Excavated in May 2008.

Table 1. List of Archaeological Reports from Cuba Mentioning Aquatic Mammals

Author	Achaeological site/ province	Period	Species name
Álvarez & Vento (1997-2001)	Canímar Abajo (Matanzas)	Proto-agricultural (3000-700 AP)**	Bottlenose dolphin *(Tursiops truncatus)*
Rey et al. (1950)*	Asiento Carapachibey (Isla de la Juventud)	Fisher-Gatherers (500 y 3 500 AP)**	Manatee (*Trichechus manatus*)
Pérez Iglesias et al. (2004)	Los Buchillones (Ciego de Ávila)	Agro-pottery (1200 CE and 1600 CE)	Bottlenose dolphin (Tursiops truncatus)
La Rosa et al. (1993)*	Tasajera I (Mayabeque)	Fisher-Gatherers (500 y 3 500 AP)**	Manatee *(Trichechus manatus)*
Rodríguez (1984)*	Cayo Carenas, (Cienfuegos)	Agro-pottery (500 y 1200 AP)**	Manatee (*Trichechus manatus*)
Herrera, (1970)*	Cayo Jorajuría (Matanzas)	Proto-agricultural (3000-700 AP)**	Manatee (*Trichechus manatus*)
Borroto & Arredondo (2011)*	Solapa Alta (Villa Clara)	Hunter (10 000-3300 AP)**	Bottle nose dolphin (*Tursiops truncatus)*
Pérez, Álvarez, Silva & Rodríguez 2007*	Toma de Agua Río Zaza, (Sancti Spiritus)	Agro-pottery (500 y 1200 AP)**	Manatee (*Trichechus manatus*)
Morales et al. (2011)	Cayo Lucas, Estero del Macío (Sancti Spiritus)	Fisher-Gatherers [500 y 3 500 AP)**	Manatee (*Trichechus manatus*)
Córdova & Arredondo (1988)*	Victoria I (Camagüey)	Fisher-Gatherers (500 y 3 500 AP)**	Manatee (*Trichechus manatus*)
Pérez Iglesias & Guarch (2000)	Cayo Bariay	Agro-pottery (1300 - 1460 CE)	Manatee (Trichechus manatus)
Rouse (1942)	Varela 3 (Holguín)	Agro-pottery (500 y 1200 AP)**	Whale or porpoise (undetermined species)
Pariente (1980); Jardines et al. (2009); Rouse (1942)**	Esterito, San Antonio, El Mango (Holguín)	Agro-pottery (500 y 1200 AP)**	Manatee *(Trichechus manatus)*

Author	Achaeological site/ province	Period	Species name
Valcárcel, Rodríguez & Pérez Iglesias (2001)	Corinthia III, (Holguín)	Proto-agricultural (3000-700 AP)**	Manatee *(Trichechus manatus)*
Yero et al. (2003); Pino (1988)	Corrales del Ojo de Toro, (Gramma)	Agro-pottery (500 y 1200 AP)**	Manatee *(Trichechus manatus)*
Morales et al. (2011)	Asiento el Carnero, (Gramma)	Fisher-Gatherers (3000-700 AP))**	Manatee *(Trichechus manatus)*
Navarrete (1981)*	Caimanes III (Santiago de Cuba)	Proto-agricultural (3000-700 AP)**	Manatee *(Trichechus manatus)*
Navarrete (1981)*	Caimanes II, Cayo Caimanes, Cayo Damas, (Santiago de Cuba)	Agro-pottery (500 y 3 500 AP)**	Manatee *(Trichechus manatus)*
Ortega & Córdova (1991)*	Abrigo Sílex (La Habana)	Fisher-Gatherers (500 y 3 500 AP)**	Monk seal (*Monachus tropicalis)*
Rives & Febles (1988); Pino (1991)**	Seboruco (Holguín)	Hunter (10 000-3300 AP)**	Monk seal (*Monachus tropicalis)*
Martínez et al. (1993)* ; Vento Franko (1997)	Cueva Pluma; Río Chico, (Matanzas)	Fisher-Gatherers (500 y 3 500 AP)**	Monk seal (*Monachus tropicalis)*

*Taken from Morales D., Torres L, Córdova A & Hernández I. (2011). Catálogo de fauna de sitios arqueológicos aborígenes de Cuba. Instituto Cubano de Antropología (2011): Informe Final de Proyecto. ICAN Archive. (unpublished).

**Chronology of Cuban aboriginal communities taken from Guarch, 1990.

For its part, the Caribbean monk seal (*Monachus tropicalis*, Gray 1850), belonging to the Order *Carnivora*, Family *Phocidae*, is a species that has now officially been declared extinct. It was formerly found across a large area encompassing the Bahamas, the Florida Keys, the Gulf of Mexico and the Caribbean Sea. It was most active at dawn and dusk and its diet was based on fish, crustaceans and mollusks (Davies 2012). This species appears even less frequently in the Cuban archaeological record than *Trichechus manatus.* According to Morales et al. (2011) and Pérez et al. (2012), osseous remains of the species have been found in Havana at the site known as Abrigo Sílex, in Matanzas at Cueva de la Pluma and in Holguín at Seboruco. In this last site, in 1980, Milton Pino found remains in strata corresponding to the paleoindians, the oldest culture found in Cuba, who are believed to have inhabited the island 5,000 years ago (Rives & Febles, 1988).

Another interesting finding associated with the Caribbean monk seal was uncovered in 1977 in Cueva Pluma in Matanzas Province: the skeleton of an aboriginal person adorned with a necklace from which hung two incisors subsequently identified by the paleontologist Oscar Arredondo as belonging to *Monachus tropicalis*. For the same period, a necklace composed of over 4,000 pieces fashioned from the teeth of mammals – including some belonging to the Caribbean monk seal – has been discovered in the Dominican Republic (Vento Franco 2001, Rivero, 1981). Vento Franco also reports that, 21 years later, in October 1998, a fragment of an adult *Monachus tropicalis* rib was found in Matanzas Province at the Río Chico site in the Yumurí Valley, 5 km from Matanzas, (Vento Franko 2001).

There is some information available on the use of *Monachus tropicalis* as a foodstuff and source of oil in the post-Columbian period, which suggests that, like *Trichechus manatus*, it was used more frequently in the pre-Colombian era than has been supposed. The archaeological evidence suggests that it was principally utilized to make ornamental objects; it is not at present certain whether it was used as a food source (Arredondo 2008).

The bottlenose dolphin (*Tursiops truncatus*, Montagu 1821) of the Order *Cetacea*, Family *Delphinidae*, is widely distributed across the tropical and temperate zones, being absent only from the Arctic and Antarctic Oceans. It is usually found in pods of between 10 and 12 individuals and feeds, in coastal areas, on a range of fish and invertebrates. It swims at speeds of between 5 and 11 km/h.

In Cuba, remains of the bottlenose dolphin in archaeological sites are as scarce as *Monachus tropicalis*. They have been found in Pinar del Río at the site known as El Cocuyo (Álvarez & Vento 1997-2001); Matanzas, in Canímar Abajo (Álvarez & Vento 1997-2001); Villa Clara, in Solapa Alta (Borroto & Arredondo 2011) and in Ciego de Ávila at Los Buchillones (Pérez Iglesias et al. 2004).

The sites of El Cocuyo, in Pinar del Río, and Canímar Abajo in Matanzas (Álvarez & Vento 1997-2001), are interesting because *Tursiops truncatus* is associated with human burials at both. The first was found by Enrique Alonso and Hilario Carmenate in 1979 and consisted of a secondary burial accompanied by a dolphin skull and shells of the species *Strombus gigas* (queen conch), *Isognomon alatus* (flat tree oyster) and *Codakia orbicularis* (tiger lucine) (Academia de Ciencias, 1988). The second find occurred during an excavation of the grave of a child aged about 12, carried out in 1996 at a stratigraphic level of 70-80 cm by the Matanzas Speleological Committee under the direction of Ercilio Vento. The rib of a dolphin, 29 cm long, was found 25 cm from the child's skull, the first find of its kind reported for Matanzas Province (Vento 1997-2001; Alonso, et al. 1988).

Marine mammals as raw material for aboriginal artistic expression

As has been noted, a significant amount of the available archaeological evidence on the use of marine mammals is connected with the elaboration of artefacts. It is therefore important to examine museum collections of ceremonial objects and body ornaments for further information on the use of marine mammals in prehistory. In the specific case of Cuba this means looking at objects from sites associated with agro-potter groups (Tabío 1984), whose inhabitants elaborated ornaments from a wide range of materials including bone, for which marine mammals were an important source because of their size.

The Banes region in Holguín Province is rich in archaeological sites of this type, some 70 being found there. Valcárcel (1999) provides information on body ornaments and their ceremonial use in an article on hierarchical relations in indigenous communities. Of an analysis of 530 pieces 9.2 % were found to have been made of bone.

Valcárcel, et al. (2003) produced a digital catalog of available written and visual information on ceremonial objects and body ornaments for the same zone. The catalog brings together a total of 200 objects of which 25 were described as being made from manatee bones (*Trichechus manatus*). These objects included vomit spatulas associated with purging rituals carried out by *behíques* or shamans in ceremonies for the invocation of the gods. Other pieces include portable idols, sets of teeth for incrusting in wood, and earrings. The following section describes some of these ceremonial objects from the Banes region[1].

Figure 3, from the Potrero del Mango habitation site. This is a long piece with a cephalic decoration at one end. The carving is anthropomorphic, with large round perforated eyes and an oval mouth, also recessed. A small peduncle appears on the back, at the level of the neck, with a biconical perforation that was probably used to hang the piece. The blade extends from the figure and is fractured very close to the far end. The object is somewhat yellowed. Height: 5.20 cm, width: 2.50 cm, depth: 1.05 cm.

[1] The descriptions and photographs are taken from Valcárcel et al., (2003).

Figure 4, from the Los Cateyes del antiguo barrio Los Jagüeyes habitation site. A long object with a cephalic image in the middle. The head is anthropomorphic, with very simple features. It has a polished surface and is yellow in color. Height: 8.60 cm, width: 2.80 cm, depth: 2.60.

Figure 5, from the Potrero del Mango habitation site. A piece with a possible cephalic representation in which two large holes reminiscent of eyes have been pierced in flat disks. The front is polished. It is a dark yellow. Height: 6.90 cm, width: 1.80 cm, depth: 1.30 cm.

Figure 3. Vomit spatula with anthropomorphic representation, from the Banes region, Holguín, Cuba (Catalog No. 5-6, Baní Museum).

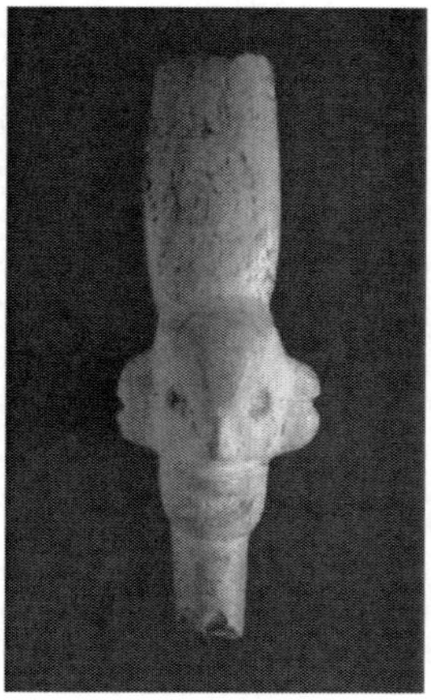

Figure 4. Vomit spatula with cephalic image in the center, from the Banes region, Holguín, Cuba (Catalog No. 5-5, Baní Museum).

Figure 5. Vomit spatula with a cephalic representation at one end, from the Banes region, Holguín, Cuba (Catalog No. 5-7 Baní Museum).

Figure 6, highly worked vomitory spatula with a zoomorphic representation showing the head of a bird with large round eyes on either side of the head. The beak, which has a small cylindrical element decorated with incisions, touches the body. The figure has a humanoid appearance suggested by two pierced disks that resemble shoulders, from which small extremities resembling feet protrude. These are separated from the body by a perforation on either side. The back displays, in the center, a large projection reminiscent of a spinal column, from which a small perforated peduncle projects, presumably used to hang the piece. The surface of the object has been polished and is a dark yellow color. Height: 6.20 cm, width: 3.20 cm, depth: 1.80 cm. Place of origin unknown.

Figure 7, from the Esterito archaeological site, a portable idol with a magical-religious function probably linked to fertility. It is an upright anthropomorphic female figure with arms by its sides, separated from the body by long perforations. The piece has lost its head from the base of the neck upwards. It is a large piece of solid appearance and is a pale yellow color. Height: 12.60 cm, width: 6.50 cm, depth: 3.50 cm.

The first four objects appear to have been fashioned from rib bones, whose size, shape, strength and length before working are similar to those of these blade-shaped pieces designed to be put down the throat to provoke ritual vomiting. Friar Ramón Pané (cited by Caro 1977) described ceremonies such as that of *La Cohoba* in which artifacts of this kind were used. In his book "La Cohoba", on the Dominican Republic, Caro provides a series of images of finely worked vomit spatulas including some that were made from the bones of marine mammals (Caro 1977). The piece shown in Figure 7 is of considerable size and was probably made from a shorter and thicker manatee bone, perhaps the shoulder girdle, pelvis or skull.

Figure 6. Vomit spatula representing an anthrozoomorphic figure from the Banes region, Holguín, Cuba (Catalog No. 5-1 in the Baní Museum).

Figure 7. Portable image with anthropomorphic representation, from Banes, Holguín, Cuba (Catalog No. 5-14, Baní Museum).

This collection of objects was donated to the Indo-Cuban Museum at Baní, Banes, in Holguín Province, by the collector Orencio Miguel Alonso who collected them himself in the mid-20th Century, or bought them from small farmers in the area. Information on the contexts in which they were found is lacking and with it important data that could aid in the interpretation of their possible uses.

The first historical record of the Caribbean monk seal dates to Columbus's second voyage during which, according to Kerr (1824), eight were hunted for food off the southern coast of Hispaniola (Villar et al. 1985; Baker 2008; Blanco et al. 2011; Daneri & De Santis 2001). Daneri and De Santis (2001) suggest that the Caribbean monk seal was sufficiently abundant between the 17th and 19th Centuries to be hunted commercially on a large scale for oil derived from its blubber, mainly by Europeans. The activity, though, is poorly documented.

The chronicler Gonzalo Fernández de Oviedo noted (1851: 428):

> There are many sea lions, and very large, in the seas of these Indies, both among the islands and on Tierra Firma. These are the swiftest and most rapid animals that live in the sea, (…). The sea lions leave [the water] to sleep on the land in many islands or in parts of the coast: and their sleep is so deep and heavy and they snore so loudly that they may be heard from afar; and thus often when they are sleeping they are killed at night (…)[2]

From this information it may be confirmed that the aboriginal agro-potters who made the objects itemized above lived at the same time as the Caribbean monk seal, but its remains are very scarce in Cuba. This finding supports the view of Vento Franco that some of the ceremonial objects or body ornaments might have been made using the ribs or other bones of the monk seal and not of the manatee, as is usually claimed (Vento Franco 1997). It is therefore important to examine which raw materials were used to make the objects, an activity which is usually given less priority by archaeologists than determining their function, their aesthetic characteristics, their meaning, or the techniques used in their manufacture.

Up until now most of the information on archaeological finds of marine mammals in Cuba has lacked precise contextual data that have permitted the use of the animals and of their remains to be evaluated. It appears that pieces made with teeth tend to be related to funerary contexts or were used in pendants. It is not possible to determine whether they had previously been

---

[2] This must refer to the Caribbean monk seal (*Monachus tropicalis*), Family *Phocidae* and not to the sea lion, which belongs to a different family of marine carnivores (the *Otariidae*) which are not found in the Caribbean.

consumed as a foodstuff. However, ethnographic data from the 15th and 16th Centuries offer very clear information on the consumption of marine mammals by the aboriginal population, principally manatees but also, potentially, seals.

Amongst the small number of sites which provide clear evidence of the consumption of manatees, of particular importance are Cayo Bariay and San Antonio Holguín Province. In both, remains of *Trichechus manatus* have been found in the context of food preparation alongside other animal remains and materials and structures that were used in cooking fires (Pérez & Guarch 2003; Jardines et al. 2009).

Other Reports from the Antilles

Information on the presence of marine mammal remains in middens in the Antilles is equally scarce. In their 2004 study of the flora and fauna of the Lesser Antilles and some parts of the Greater Antilles, Newsom and Wing reported the presence of the remains of two *Monachus tropicalis* for archaic sites in Nevis, specifically in Hichmans' Shell Heap, and remains of an unspecified cetacean at the Jolly Beach site on the island of Antigua. For later cultures, such as the agro-potter peoples, the same authors report the presence of *Monachus tropicalis* and dolphin (*Stenella* sp.) in Santa Bárbara, Curaçao (Newsom & Wing, 2004).

For its part, a zooarchaeological analysis of material from the Maisabel site in Puerto Rico found that from a large sample of animal remains, the rarest were marine turtles and manatees (De France et al. 1996).

Further interesting information about the representation of a whale in an aboriginal petrograph at the site in the Dominican Republic known as Cueva de la Línea, Cueva de los Muñecos or Cueva del Ferrocarril was provided by Torres et al. (2011). This site has over a thousand examples of cave art, the drawing being unique as it is the only artistic representation of a marine mammal in the entire Antillean area. Figure 8.

Cueva de la Línea is located on the southern side of the Bay of Samaná, one of the world's most important breeding sites for the majestic North Atlantic humpback whale (*Megaptera novaeangliae*, Borowski, 1781). However, despite the abundance of habitation sites dotted around the bay, none of them provides any material culture evidence of interaction between whales and human beings, other than the extraordinary case of the Cueva de la Línea pictograph (Torres et al. 2011).

In his 2004 article on *Monachus tropicalis*, Adam reports that remains of the species have been found in archaeological sites in the Antilles, such as the Taíno indian shell midden in Ponce, Puerto Rico in the Greater Antilles. In the

Lesser Antilles, they have been found at Cinnamon Bay (Saint John, U.S. Virgin Islands), Sint Eustatius (Netherlands Antilles) and at Hichmans' Site and Hichmans' Shell Heap in Nevis (Saint Christopher-Nevis). Two further cases have been reported for Santa Barbara and Sint Michiel in Curaçao.

Figure 8. Pictograph of a whale in the Cueva de la Línea, Bahía de San Lorenzo, southern shores of the Bay of Samaná, Dominican Republic.

Northern South America: The Case of Colombia

The information discussed in this section is based on earlier studies (Ramos 2008; 2013) that compiled and analyzed information on the presence of fauna in Colombian archaeological sites. The first reviewed the majority of the published archaeological research reports and papers in order, firstly, to register the animal species that have been found and with what frequency and, secondly, the kind of analysis they have been subjected to in each site. Building on the results of this research the second article (Ramos 2013) sought to evaluate and discuss the biases that might have affected the collection and interpretation of the samples.

In addition, a bibliographical search was carried out of archaeological and ethnozoological research on these species in Colombia, with particular reference to marine mammals.

Table 2 and Figure 9 present a list of the publications that have reported the presence of marine mammals in Colombia and describe the geographical location of the reported sites.

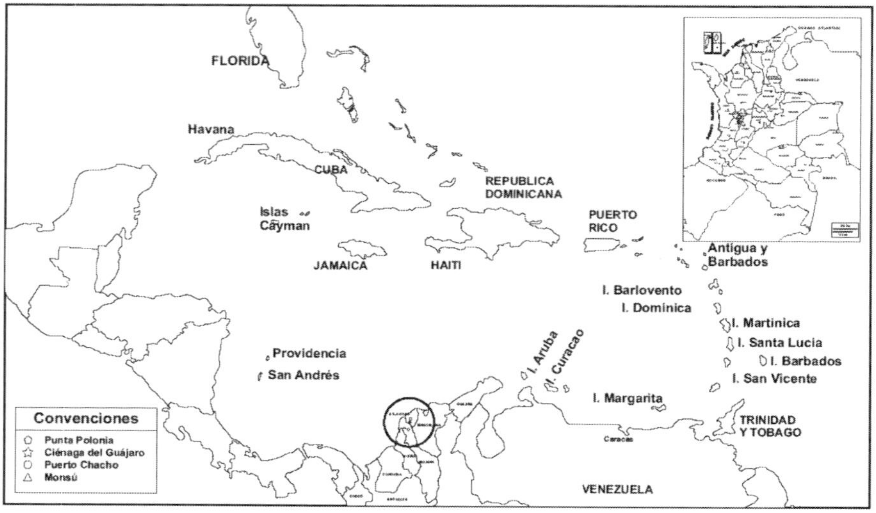

Figure 9. Localization Map of Colombian sites where marine mammal species have been reported.

Table 2. List of Archaeological Reports from Colombia Mentioning Marine Mammals

Author	Achaeological site/ department	Period	Species name
Álvarez, R. & Maldonado, H. (2009)	Puerto Chacho (Colombia)	Early formative [approx. 3300 BCE]	Manatee (*Trichechus manatus*)
Álvarez, R. & Maldonado, H. (2010)	Puerto Chacho (Colombia)	Early formative [approx. 3300 BCE]	Manatee (*Trichechus manatus*)
Angulo, C. (1987)	Guájaro	3000 BCE – 16th Century CE.	Manatee (*Trichechus manatus*)
Angulo, C. (1988)	Guájaro	Rotinet [3000 BCE-2000 BCE], Carrizal [0 CE- Conquest (Saíno Phase, Palmar Phase)]	Manatee (*Trichechus manatus*)
Betancourt, A. (2003)	Punta Polonia	Early formative (related to Rotinet)	Manatee (*Trichechus manatus*)

Author	Achaeological site/ department	Period	Species name
García, M.L. (1997)	Puerto Chacho, Monsú, San Jacinto 2, Canapote, Barlovento, El Pozon, Bucarelia, El Guamo,	Early formative	Manatee (*Trichechus manatus*)
González V. (1989)	Puerto Chacho	Early formative (Medium occupation period 3270 ± 90 BCE)	Manatee (*Trichechus manatus*)
Legros, T. (1990)	Puerto Chacho	3270 BCE (Un-calibrated date)	Manatee (*Trichechus manatus*)
Reichel-Dolmatoff, G. (1985)	Monsú	3350 ± 80 BCE - 1st to 2nd Century BCE	Manatee (*Trichechus manatus*)

This information confirms that few marine mammal remains have been found in archaeological sites in Colombia. Of the 230 studies reviewed the few that registered marine species are shown in Table 2 and represent just one species, the manatee *(Trichechus manatus)*, whose remains were found at Puerto Chacho (Alvarez & Maldonado 2009, 2010; Legros 1990; Gonzáles 1989), Punta Polonia (Betancourt 2003), Guájaro (Angulo 1987,1988) and Monsú (Reichel-Dolmatoff 1985). In addition, certain artefacts from the Sierra Nevada de Santa Marta, today in the collection of the Gold Museum in Bogotá, are believed to have been made from manatee bones (Ramos 2009:151). The prehistoric presence of manatees has also been reported for the Lower Río San Jorge (Plazas and Falchetti 1981) but without details as to the use made of them. For these sites the total number of manatee bones reported is as follows: 88 in Puerto Chacho (García 1997), 16 in Punta Polonia, (Betancourt, 2003) and four in Monsú (Reichel-Dolmatoff, 1985), while for excavations in the environs of the Embalse de Guájaro no concrete information is available.

Up until now, these are the only concrete cases that have reported the use of marine mammals in present-day Colombia during the prehistoric era. As can be seen in Table 2 these cases are early, dating between 3,350 and 2,000 BCE, during the Early Formative Period (Langebaeck & Dever 2000).

It is apparent from this body of evidence that the only archaeological site that presents concrete evidence that the manatee might have been utilized as a food source is Puerto Chacho in the municipality of Arjona, Bolívar Department, where manatee remains were the second most abundant amongst those of the fauna recovered (García 2007). It has been suggested that manatee flesh was eaten at this site only occasionally, possibly for mythical or religious reasons (Álvarez & Maldonado, 2009; 2010; García 1997). In support of the hypothesis that manatee flesh was consumed, García (1997) mentions that some bones had been exposed to fire and displayed signs of butchery or of meat extraction. This is the strongest evidence found hitherto of the consumption of marine mammals in this part of the continent. None of the other sites render specific evidence of the manatee having been consumed. Angulo (1988) reports evidence that manatee bones were used to fashion harpoons in Guájaro while for Punta Polonia Betancourt makes no mention of their possible use. Additional evidence on the utilization of the manatee for the elaboration of ornaments comes from a few examples from the Sierra Nevada de Santa Marta now in the collection of the Gold Museum in Bogotá (Ramos 2009), for which it has been established that the large or heavy bones of the manatee - such as the scapula, pelvis and parts of the cranium - were used to make artefacts including snail-shaped pendants, bone spindle weights, fittings for ceremonial staff heads and harpoons. However, although it is certain the artefacts originated in the Tairona archaeological zone no contextual information is available.

Different aspects need to be taken into account when interpreting the scarce evidence for the utilization of marine mammals in the region. First, it is significant that of the 57 archaeological sites that have been documented to contain relatively well preserved animal remains, the species has been found in only four. It may therefore be deduced that the problem is not one of preservation, especially as the bones of marine mammals are large and heavy and are more likely than other remains to survive the taphonomic process. Two principal reasons might explain their scarce representation in these sites in the Colombian Caribbean, where all the sites are located (Figure 9). The first of these involves the difficulty that the size, ecology and behavior of marine mammals could have posed for hunters, and the second, the great abundance of vertebrates from lacustrine, river and maritime ecosystems which do not produce as much meat as marine mammals but are much easier to locate and to capture. These animals include alligators, turtles, iguanas and many species of fish and terrestrial mammals, the remains of which are abundant in the archaeological sites of the region (Ramos 2010 a and b; Ramos and Archila 2008), a finding that supports the idea that these animals were important to the subsistence of human populations.

It is not possible to argue, on the basis of this scarce evidence, that marine mammals were overexploited or even, necessarily, utilized for human consumption in prehistoric times. However, the evidence is very fragmentary and investigations have not up till now been conducted to examine the hypothesis more carefully. It might be asked why species like the manatee are today practically extinct in the region if there is no evidence of high levels of consumption in the past, or at least in the pre-hispanic period. However, the information available from primary and secondary sources from the 15th and 16th Centuries, though not abundant, is clear about the utilization of the manatee. According to Saldarriaga (2006) in some zones of present-day Colombia, and particularly at the confluence of the Magdalena and Cauca rivers, the manatee became important to the economy of the peoples of the region, leading the Spanish to include it in the tribute required of them under the *encomienda* system and leading, over time, to a considerable decrease in the population of the species. Saldarriaga argues that the principal reason for its capture was not as a source of meat (though this was consumed) but for the extraction of oil (fat) from its body. This was used as fuel for lamps and also in food preparation. (For a detailed review of the historical sources on this matter, see Carvajal in this volume). Today, according to Álvarez and Maldonado (2010:117), the manatee has been declared an endangered species in the Colombian Amazon, Orinoco region and Caribbean because "its meat is appreciated for human consumption and because of illegal activity, the use of fishing nets, the drying-up of wetlands (*ciénagas*) and the pressure caused by boats with outboard motors".

CONCLUSION

On the basis of the information on the Antilles and northern South America discussed in this chapter (Table 3) it may be concluded, for Colombia, that though marine mammals might have constituted an important food source and significant raw material for artefacts, they did not in any case represent the largest percentage of the fauna found on the archaeological sites, their use apparently being limited to occasional captures. For Cuba and other islands of the Greater Antilles the archaeological evidence suggests a similar conclusion, though for Colombia, ethnographic sources on these species for the 15th and 16th Centuries provide contrasting accounts.

A review of these sources uncovered the following references to the manatee:

Las Casas (1875), reports Christopher Columbus's notes on his journey along the Cuban coast on 29 October 1492:

> …and he says there were cows there and other livestock because he saw bone heads there that seemed to be of cows…that these were probably manatee, a very big fish, as big as a bullock, whose leather does not flake, like that of whale, and whose head is almost like that of a cow…

Oviedo (1851) says:

> "Concerning these manatees some are so large that they measure fourteen or fifteen feet in length and more than eight palms wide (…). They have just two hands or arms, close to the head, and for this reason the Christians called them manatees [lat.: manatus = having hands] (…)"

For his part, Francisco López de Gómara (1941) notes:

> …Eating manatee it seems more like meat than fish; fresh, it tastes of veal and salted, of tuna, though it is better and keeps well; the lard they extract from it is very good and does not go rancid; they tan their hides with it and it is used for shoes and other things; it has certain stones in its head which are used to treat kidney stones and the lower abdomen … they tend to kill them when they are grazing on the river banks and, when they are small, with nets. …

Other accounts exist for the Antilles that magnify the role of species like the manatee. Mártir de Anglería, quoted by Arrom (Ulloa & Corbea 2011), refers to the capture of a manatee calf by an Indian called Caramatex:

> … and he looked after it for twenty and six years in a lake called Guainabo, where it lived; it was so sensitive, though large, and so gentle and friendly, that it eclipsed what the dolphins were for the ancients; it ate from their hands whatever it was given; it came when they called its name, Mato, which means magnificent; it left the water to eat in the house; it frolicked on the banks with the boys and the men; it showed its delight when they sang; it allowed them to climb on its back and took men from one side of the lake to the other without throwing them off, taking ten at a time with no difficulty; and this way the indians had a wonderful time with it...

In his reflections on the species Arrom notes that the indigenous population recognized that the manatee was not a fish because, despite its aquatic habits it had breasts and suckled its young (Ulloa & Corbea, 2011).

These observations by the chroniclers and other investigators illustrate the significance that the manatee had for the indigenous communities of the region and suggest that it was utilized a lot, while the archaeological record does not support this view.

It is also important to reflect on how difficult it must have been for humans to capture other marine mammals, given their large size, the absence of appropriate hunting technologies and the fact they could move rapidly in a medium in which they were much more comfortable than were humans (Guarch 1978). One aspect that supports this position is the greater abundance of *Trichechus manatus* remains in the archaeological record, as it was easier to gain access to them than to other marine mammals because of their dietary habits, their sluggish movements and the fact that they lived in calm places such as estuaries.

Table 3. List of Archaeological Reports from Colombia and the Antilles Mentioning Marine Mammals

Achaeological site/department	Author	Period[1]	Species name
Puerto Chacho (Colombia)	Álvarez, R. & Maldonado, H. (2009)	Early formative [approx. 3300 BCE] [approx. 3300 BCE]	(*Trichechus manatus*)
Puerto Chacho (Colombia)	Álvarez, R. & Maldonado, H. (2010)	Early formative [approx. 3300 BCE] [approx. 3300 BCE]	(*Trichechus manatus*)
Guájaro (Colombia)	Angulo, C. (1987)	3000 A.C - XVI d. C.**	(*Trichechus manatus*)
Guájaro (Colombia)	Angulo, C. (1988)	Rotinet [3000 BCE-2000 BCE], Carrizal [0 CE- Conquest (Saíno Phase, Palmar Phase)]	(*Trichechus manatus*)
Puerto Chacho (Colombia)	Arévalo, H. & Maldonado, H. (1990)	Early formative [approx. 3300 BCE]	(*Trichechus manatus*)
Punta Polonia (Colombia)	Betancourt, A. (2003)	Early formative [approx. 3300 BCE] (associated with Rotinet)	(*Trichechus manatus)*
Maisabel (Puerto Rico)	De France et al.(1996)	Ceramic Age 700–510 BP	*(Trichechus manatus)*
Caracoles Midden (Ponce, Puerto Rico)	Narganes, pers. comm.; Wing 1992	Ceramic Age 700–510 BP	(*Monachus tropicalis)*
Cinnamon Bay (Saint John, U.S. Virgin Islands)	Quitmyer and K. Wild*****	Archaic period 1,000–510 BP	*(Monachus tropicalis)*

Table 3. (Continued)

Achaeological site/department	Author	Period[1]	Species name
Sint Eustatius (Netherlands Antilles)	van der Klift 1992*****	Archaic period 1,400–1,100 BP**	*(Monachus tropicalis)*
Hichman's Site Nevis (Saint Christopher-Nevis)	Wing 1992, 2001*****	Archaic period 1,660 BP	*(Monachus tropicalis)*
Hichman's Shell Heap Nevis (Saint Christopher-Nevis)	Wing 1992, 2001*****	Archaic period 2,550 BP	*(Monachus tropicalis)*
Santa Bárbara Curazao	Haviser 1994*****	Ceramic Age, 400–1,600 BP	*(Monachus tropicalis)*
Sint Michiel	Debrot 2000; Haviser 1987*****	Archaic Age, 3,790–3,820 BP	*(Monachus tropicalis)*
Puerto Chacho (Colombia)	González V. (1989)	Early formative [approx. 3300 BCE] (Ocupación media 3270 ± 90 a.C)	(*Trichechus manatus*)
Puerto Chacho (Colombia)	Legros, T. (1990)	3270 a.C (Fecha no calibrada)	(*Trichechus manatus*)
Hichman's Shell HeBP (Nevis)	Newsom, L.A. & Wing, E.S. (2004)	Archaic period*	(*Monachus tropicalis)*
Jolly Beach, Antigua	Newsom, L.A. & Wing, E.S. (2004)	Archaic period*	Whale or porpoise
Canímar Abajo, Matanzas (Cuba)	Álvarez &Vento (1997-2001)	Proto-agricultural**** (3000-700 AP)	*(Tursiops truncatus)*
Los Buchillones, Ciego de Ávila (Cuba)	Pérez Iglesias et al.(2004)	Agro-pottery (1200 CE and 1600 CE)	*(Tursiops truncatus)*
Monsú (Colombia)	Reichel-Dolmatoff, G. (1985)	3350 ± 80 BCE - I a II a.C**	(*Trichechus manatus*)
Varela 3, Holguín (Cuba)	Rouse (1942)	Agro-pottery (500 y 1200 AP)****	Whale or porpoise
Tasajera I (Cuba)	La Rosa et al. (1993)***	Fisher-Gatherers (500 y 3 500 AP)**	*(Trichechus manatus)*
Asiento Carapachibey (Isla de la Juventud, Cuba)	Rey et al.(1950)***	Proto-agricultural** (3000-700 AP)	*(Trichechus manatus)*
Cayo Jorajuría (Matanzas, Cuba)	Herrera, (1970)***	Proto-agricultural (3000-700 AP)****	*(Trichechus manatus)*

Achaeological site/department	Author	Period[1]	Species name
Cayo Carenas (Cienfuegos, Cuba)	Rodríguez, (1984)***	Agro-pottery (500 y 1200 AP)****	(*Trichechus manatus*)
Solapa Alta (Villa Clara, Cuba)	Borroto y Arredondo, (2011)***	Hunters (10 000-3300 AP)****	(*Tursiops truncatus*)
Toma de Agua Río Zaza (Sancti Spiritus, Cuba)	Pérez, Álvarez, Silva y Rodríguez 2007***	Agro-pottery (500 y 1200 AP)****	(*Trichechus manatus*)
Cayo Lucas, Estero del Macío (Sancti Spiritus, Cuba)	Morales et al., 2011	Proto-agricultural (3000-700 AP)****	(*Trichechus manatus*)
Victoria I (Camagüey, Cuba)	Córdova y Arredondo, 1988***	Fisher-Gatherers (500 y 3 500 AP)****	(*Trichechus manatus*)
Cayo Bariay, Esterito, San Antonio, El Mango (Holguín, Cuba)	Pérez Iglesias y E Guarch, 2000; Pariente, 1980; Jardines et al., 2009; Rouse, 1942***	Agro-pottery [500 y 1200 AP]****	(*Trichechus manatus*)
Corinthia III (Holguín, Cuba)	Valcárcel, Rodríguez y Pérez Iglesias, 2001	Proto-agricultural (500 y 3 500 AP)****	(*Trichechus manatus*)
Corrales del Ojo de Toro (Gramma, Cuba)	Yero et al., 2003, Pino, 1988	Agro-pottery (500 y 1200 AP)****	(*Trichechus manatus*)
Asiento el Carnero (Gramma, Cuba)	Morales et al., 2011	Fisher-Gatherers (500 y 3 500 AP)****	(*Trichechus manatus*)
Caimanes III (Santiago de Cuba, Cuba)	Navarrete, 1981***	Proto-agricultural (3000-700 AP)****	(*Trichechus manatus*)
Caimanes II, Cayo Caimanes, Cayo Damas (Santiago de Cuba, Cuba)	Navarrete, 1981***	Proto-agricultural (3000-700 AP)****	(*Trichechus manatus*)
Abrigo Sílex (La Habana, Cuba)	Ortega y Córdova, 1991***	Fisher-Gatherers (500 y 3 500 AP)****	(*Monachus tropicalis*)
Seboruco (Holguín, Cuba)	Rives y Febles, 1988 (Pino, 1991)***	Hunters (10 000-3300 AP)****	(*Monachus tropicalis*)
Cueva Pluma; Río Chico (Matanzas, Cuba)	Martínez et al., 1993*** ; Vento Franko, 1997	Fisher-Gatherers (500 y 3 500 AP)****	(*Monachus tropicalis*)

[1] The dates appearing in the period column are presented as BP or BCE according to the usage of the authors of the works referred to.

* No precise date for the remains is given by authors of the works referred to.

** Neither the period nor the date of the remains is provided in the works referred to.

*** Taken from Morales D., Torres L, Córdova A & Hernández I. (2011). Catálogo de fauna de sitios arqueológicos aborígenes de Cuba. Instituto Cubano de Antropología (2011): Informe Final de Proyecto. ICAN Archives. (Unpublished).

****As no absolute chronologies have been established, the dates provided here refer to the periods when Guarch (1990) calculates the communities were inhabitted.

***** Taken from Adam P. J. MAMMALIAN SPECIES. Monachus tropicalis. In American Society of Mammalogists. No. 747. (2004)

As has been seen, the marine mammal that is most abundantly represented in the archaeological sites of the Antilles and Colombia is the manatee, while other mammals, such as dolphins, whales and seals (including the now-extinct *Monachus tropicalis*) appear only occasionally. This could respond to the fact that, in contrast to the manatee, these other marine mammals are rapid movers, a factor which – combined with the relative complexity of the technology required to catch them (Guarch 1978) - might in part explain their scarcity in the archaeological record. An element that is common to the evidence from all the sites examined is that it is not possible to establish unequivocally whether these species were used as a source of food.

The scarcity of available information on the presence of marine mammals is common to the whole of the Antilles. In their analysis Newsom and Wing (2004) argue that inland groups might have exploited them, using strategies that involved butchering and discarding of major skeletal remains at the coast and then transporting cuts of meat and soft parts to their settlements where they were consumed (Newsom & Wing 2004) and (Jiménez & Arredondo 2011).

Despite their scarcity in the record, it must be the case that the capture of some of these species was important because of the large quantity of biomass they provided and because, in particular for agro-pottery groups, they were highly valued for the raw material their bones provided for the elaboration of offerings, body ornaments and as elements of ceremonial paraphernalia. As has been seen, the Banes region, in Holguín Province, Cuba (some of whose sites are considered to be inland at between 8 and 10 km from the coast) provides numerous ceremonial objects and ornaments. Obtaining the raw materials needed to make these artifacts must have constituted a significant activity, requiring inhabitants to make the effort to locate the animal or to engage in some form of exchange activity.

The pictograph of a whale in Cueva de la Línea in the Dominican Republic is an exceptional depiction of a marine mammal. It has been suggested that the drawing had an anecdotal purpose, allowing knowledge to be passed between inhabitants or individuals engaged in magic-religious processes associated with the direct effect whales could have on the nutrition and daily life of the community (Torres et al. 2011).

The use of marine mammals by humans has also been identified in other regions of the continent. Cetaceans (dolphins and whales) have been found in sites in Argentina, Chile, Brazil, Mexico, the Dutch Antilles and on the island of Antigua (Volkmer de Castilho & Simões-Lopes 2008; Götz & Sierra 2011; Newson & Wing, 2004). In terms of the specific use of these species it is inferred that in Brazil they were eaten, though they were not the principal food source for communities (Volkmer de Castilho & Simões-Lopes 2008). Once again, the commonest species is the manatee, which has been found in Central American countries including Belize, Panama, Honduras, Nicaragua and Guatemala (Emery 2004), among which it is commonest in the archaeological records of Panama and Belize. In Panama manatees have been reported for the sites known as Sitio Conté, Bocas del Toro, Bahía de Parita, Cerro Mangote and el Hatillo (Cooke & Martín-Rincón 2010; Cooke & Sánchez 2001; Cooke & Sánchez 2004a; 2004b; Cooke et al. 2007; McKillop, 1985). It has also been inferred that these manatee remains were exchanged or consumed and that artefacts made from them had a "religious significance or were symbols of power" (Cooke & Sánchez, 2004b:578) and were thus associated with elites.

In Belize, evidence has also been found that the meat of manatees was consumed in different communities and that their bones were used to make small carvings or sculptures, needles, musical instruments (made from ribs) and other ceremonial artefacts (McKillop 1985; Masson 2004). It has also been suggested that the manatee was consumed in Nicaragua (Cooke & Martín-Rincón 2010; McKillop 1985), and that some of its remains might have had symbolic power, being kept as good luck talismans (McKillop 1985). One important piece of information that helps to illustrate the treatment given to fauna, and the difficulty of identifying marine mammal remains in the archaeological record, is provided by a case reported by McKillop (1985) for Nicaragua, where the Rama indigenous community returned manatee remains to the sea, as they believed this would bring luck to later hunting parties. Additionally, Henderson and Joyce (2004), reported for the islands of St. Kitts, Grenada and Trinidad (McKillop 1985), while for Guatemala its presence is reported but there is no evidence of how it was used.

The remains of three other marine mammal species (sea lion: Family *Otariidae*; Subantarctic fur seal: *Arctocephalus tropicalis*; and South American sea lion: *Otaria flavescens*) have been reported for Brazil (Volkmer de Castilho & Simões-Lopes, 2008), while remains of the extinct *Monachus tropicalis* have been found in Mexico, Nevis and Curaçao, (Newsom & Wing 2004, Götz & Sierra 2011).

It is very interesting to contrast McKillop's conclusions (1985: 348) on the exploitation of the manatee in Mayan territory and the circum-Caribbean area and the evidence uncovered for Cuba, the Antilles and Colombia in terms of the significance of marine mammal use for aboriginal populations:

> "For the Maya and circum-Caribbean regions marine animals have been more important than land animals to the diet at those sites where reliable recovery of significant quantities of meat and/or fish could be obtained. Such a reliable recovery depended on various factors including travel distance to pray [sic.], characteristics of the available animal species (such as size, grouping, mobility, and abundance), labor costs, the practice of live animal tending or meat and fish preservation, the nature of other subsistence activities, as well as access to an external economy".

The currently available evidence suggests that, at least for the particular case of northern Colombia and the Antilles, this was not the case with the manatee and other marine mammal species. On the contrary, zooarchaeological studies in these places suggest that other vertebrates such as fish, reptiles and terrestrial mammals were the principal food sources for these communities.

To summarize, in order to be able to respond to the research questions posed at the start of the chapter on the use of marine mammals in the Antillean region and northern South America (Colombia), and in order to be able to argue, as has this chapter has, that the use of marine mammals in the region was not so systematic, it is necessary to evaluate the available evidence more carefully and to develop methodological strategies that permit archaeofaunal complexes to be reconstructed. It is, thus, a priority to examine in greater detail the taphonomic factors that might have had a significant effect on the appearance of marine mammals in the sites reported. During excavations it is of particular importance to evaluate the cultural factors associated with the strategies by which marine mammals were selected, captured, processed and parts of them discarded. It is well known that these factors might vary significantly between different cultural groups. We believe, therefore, that focused studies should be carried out in the different regions, with a significant ethnobiological component, similar to those that have focused on the manatee in the Brazilian Amazon (Calvimontes & Marmontel 2010). The information provided by such ethnobiological studies might be of great significance in understanding the use these animal species were put to and might also inform programs for animal conservation and sustainable development as well as the preservation of the cultural traditions of the peoples of the region.

Acknowledgments

We wish to thank Dany Morales MSc. Of the Institute of Anthropology (ICAN) in Havana for the data used to complete the zooarchaeological register compiled for this chapter; Pedro Luis Hernández of the Speleological Committee of Pinar del Río for his collaboration concerning information reported on the remains of marine mammals in sites in Pinar del Río; Laura de Luca, a Masters student in the Department of Anthropology at the Universidad de los Andes in Bogotá for her collaboration compiling the references for Colombia; and the archaeologists Luis Gonzalo Jaramillo and Roberto Valcárcel Rojas for their comments on the manuscript. Finally, we are grateful to the Social Science Faculty at the Universidad de los Andes for making funds available for the translation of the article and to James Lupton and Tiziana Laudato for preparing the translation and for their valuable comments. All errors remain the responsibility of the authors.

References

Adam P. J. (2004). Mammalian species. Monachus tropicalis. *American Society of Mammalogists*, *747*, 1–9.

Alonso, E., Carmenate, H. & Blanco, E (1988). Mapa Arqueológico de Pinar del Río [Mimeographed]. Havana: Academia de Ciencias de Cuba. Delegación territorial Pinar del Río.

Álvarez, A. (2012). Trichechus manatus. In H. González, L. Rodríguez, A. Rodríguez, C.A. Mancina & I. Ramos García (Eds.). *Libro Rojo de los Vertebrados de Cuba* (pp. 289-291). Havana, Cuba: Editorial Academia.

Álvarez, A. & Vento, E. (2001). El delfín y su probable vinculación como ofrenda en entierros aborígenes de Cuba. 1861 *Revista de Arqueología y*

Espeleología. [On-line serial]. No 1 – 4, Available at [http://www.italia-cuba.speleo.it/1861_1-4.pdf]

Álvarez, R. & Maldonado, H. (2009). Arqueofauna encontrada en Puerto Chacho, sitio arqueológico del Caribe colombiano (3300 a.C.). *Revista de la Academia Colombiana de Ciencias Exactas, Físicas y Naturales, 33 (128)*, 407-424.

Álvarez, R. & Maldonado, H. (2010). El manatí caribeño Trichechus manatus Linnaeus, 1758, en los restos faunísticos de Puerto Chacho (3300 A.C.), Caribe colombiano. *Boletín Científico Centro de Museos Museo de Historia Natural, 14 (2)*, 101-119.

Angulo, C. (1987). *Guájaro: Un modo de vida en la arqueología del Norte de Colombia*. Anuario Científico VI. Barranquilla: Ediciones Universidad del Norte.

Angulo, C. (1988). *Guájaro en la arqueología del Norte de Colombia*. Primera Edición. Bogotá, Colombia: Fundación de Investigaciones Arqueológicas Nacionales Banco de la República.

Arévalo, H. & Maldonado, H. (1990). Una propuesta al formativo temprano en Colombia. Undergraduate thesis. Universidad Nacional de Colombia, Bogotá, Colombia.

Arredondo, C. (2008). Arqueozoología prehispánica en Cuba: situación actual y desarrollo. In G. Mengoni Goñalons, J. Arroyo-Cabrales & O. J. Polaco (Eds.). *Latin-American Archaeozoology: Origins and Development* (pp. 153-163). Mexico DF: Instituto Nacional de Antropología e Historia y Consejo Nacional para la Ciencia y la Tecnología.

Baker, K. (2008). Tantas focas en tan poco tiempo. La rápida extinción de la foca monje del Caribe. *El Monachus Guardián, 11, (1)*, June. Available at [http://www.monachus-guardian.org/spanish/mguard21es/2131scien.htm]

Betancourt, A. (2003). Punta Polonia y el formativo temprano en Colombia. Unpublished undergraduate thesis, Universidad Nacional de Colombia.

Blanco, M., Borroto-Páez, M. & Mancina, C. A. (2011). La foca extinta. In M. Borroto & C. A. Mancina (Eds.). *Mamíferos en Cuba*. (pp. 202-205). Vasa: Upc Print.

Calvimontes, J. & Marmotel, M. (2010) (eds.). Estudios etnobiológicos sobre el manatí amazónico (Trichechus inunguis naterer) y su conservación en la reserva de desarrollo sostenible Amaná, Brasil. In *Sistemas Biocognitivos tradicionales. Paradigmas en la conservación Biológica y el Fortalecimiento cultural* (pp. 397-402). Mexico D.F: Asociación Etnobiológica mexicana, Global Diversity Foundation, Mexico.

Universidad Autónoma de Hidalgo, Colegio de la Frontera Sur land the Sociedad Latinoamericana de Etnobiología.

Caro, J. A. (1977). *La Cohoba*. Barcelona, España: Artes Gráficas Manuel Pareja.

Cooke, R. G. & Sánchez, L. (2001). El papel del mar y de las costas en el Panamá Pre-hispánico y el periodo del contacto: redes locales y relaciones externas. *Revista de Historia*, *42*, 15-60.

Cooke, R. G. & Sánchez, L. A. (2004a). *Historia General de Panamá, Primera Parte: Las Sociedades Originarias, vol. I, T. II*. Panama City, Panamá: Comité Nacional del Centenario de la República de Panamá.

Cooke, R. G. & Sánchez, L. A. (2004b). Arqueología en Panamá (1888-2003). In *Panamá: Cien años de República*. Panamá: MANFER, S.A

Cooke R. G, Jiménez, M. & Ranere, A. (2007). Influencias humanas sobre la vegetación y fauna de vertebrados de Panamá: Actualización de datos arqueozoológicos y su relación con el paisaje antrópico durante la época precolombina. In E.C. Leigh Jr., A. Herre, J. Jackson & F. Santos-Granero (Eds.). *Ecología y Evolución en los Trópicos* (pp. 562-593). Panama City: Editora Novo Art. Smithsonian Tropical Research Institute.

Cooke, R. G. & Martín-Rincón, J. G. (2010). Arqueozoología en la Baja América Central (Nicaragua, Costa Rica, Panamá). In G. Mengoni, J. Arroyo-Cabrales, O.J. Polaco & F.J. Aguilar (Eds.). *Estado Actual de la Arqueozoología Latinoamericana/Current Advances for the Latin-American Archaeozoology* (pp. 105-131). México D.F: Instituto Nacional de Antropología e Historia: Consejo Nacional de Ciencia y Tecnología. International Council for Archaeozoology. Buenos Aires: Universidad de Buenos Aires.

Daneri, G. & De Santis, L. J. M. (2001). Distribución, historia natural y conservación de mamíferos neotropicales. *Mastozoología Neotropical. Neotrop. Mammal*, *9*, *(1)*. 59-63 Available at [www.sarem.org.ar/mastozoologia-neotropical-vol9-no1/&#8206].

Davies, L. (2012). Monachus tropicali*s*. Animal Diversity Web. University of Michigan, Museum of Zoology. Available at [http://www.ummz.lsa.umich.edu].

De France, S. W., Keegan, W. & Newsom, L. (1996). The archaeobotanical, bone isotope, and zooarchaeological records from Caribbean sites in comparative perspective. In E. Reitz, L. Newsom & S. Scudder (Eds.), *Case Studies in Environmental Archaeology* (pp. 289-304). New York, USA: Springer.

Diaz-Franco, S. (2011). Los Perezosos extintos de Cuba: Algunas reflexiones a propósito de los últimos fechados radiocarbónicos. *El Caribe Arqueológico*, *12*, 15-29.

Emery, K. F. (2004). Environments of the Maya Collapse: A Zooarchaeological Perspective from the Petexbatún. In K.F. Emery (Ed.). *Maya Zooarchaeology: New Directions in Method and Theory* (pp. 81-96). Los Ángeles, USA: Cotsen Institute of Archaeology.

García, M. L. (1997). Zooarqueología del formativo temprano de la Costa Caribe: Un acercamiento a la dieta del grupo prehistórico de Puerto Chacho. Unpublished undergraduate thesis, Universidad Nacional de Colombia.

Guarch, J. M. (1978): *El Taíno de Cuba*. Havana, Cuba: Editorial Academia.

Gónzales, V. (1989). Estudio de la Economía de Conchero de la Costa Norte: Puerto Chacho visto como un sistema adaptativo a nivel regional. Bogotá, Colombia. Department Anthropology, Unpublished undergraduate thesis, Universidad Nacional de Colombia.

Götz, C. M. & Sierra, T. M. (2011). La Arqueofauna de Xcambó, Yucatán, México. *Antípoda Revista de Antropología y Arqueología*, *13*, 119-145.

Henderson, J. S. & Joyce, R. A. (2004). Human Use of Animals in Prehispanic Honduras: A Preliminary Report from the Lower Ulúa Valley. In K.F. Emery (Ed.). *Maya Zooarchaeology: New Directions in Method and Theory* (pp. 223-238). Los Angeles, USA: Cotsen Institute in Archaeology.

Jardines, J. E, Pérez, L., Guarch, E., Cruz, P. & Martínez, M. (2009). San Antonio, sitio aborigen de contacto indohispánico de la costa este de Gibara, Holguín, Cuba. *Revista de Historia de Holguín*. Registro Nac. de Public. Seriadas: No. 2145- Folio 115- Tomo III. Septiembre - Diciembre 2009. Available at [www.baibrama.cult.cu/instituciones/rhistoria/delay/.../04.httm].

Jiménez, O. & Arredondo, C. (2011). Los mamíferos en la Arqueozoología. En R. Borroto-Páez & C. A. Mancina (Eds.). *Mamíferos en Cuba* (pp. 206-211). Finland: PC Print.

Las Casas, B. (1875). *Historia de las Indias. Tomos I y II*. Madrid, España: Imprenta de Miguel Gisnesta.

Langebaek, C. & Dever, A. (2000). *Arqueología en el Bajo Magdalena: Un estudio de los primeros agricultores del Caribe Colombiano. Informes Arqueológicos del Instituto Colombiano de Antropología e Historia No. 1*. Bogotá, Colombia: Instituto Colombiano de Antropología e Historia y Universidad de los Andes.

Legros, T. (1990). Consideraciones sobre Puerto Chacho: Un conchero de las llanuras del Caribe colombiano. *Memorias del Simposio de Arqueología y Antropología Física, V Congreso Nacional de Antropología* (67-78). Villa de Leiva, Colombia: ICFES.

López de Gómara, F. (1941). *Historia general de las Indias*. Madrid, España: Espasa-Calpe, S. A.

Masson, M. A. (2004). Fauna Exploitation from the Preclassic to the Postclassic periods at Four Maya Settlements in Northern Belize. In K.F. Emery (Ed.). *Maya Zooarchaeology: New Directions in Method and Theory* (pp. 97-122). Los Angeles, California: Cotsen Institute of Archaeology.

McKillop, H. I. (1985). Prehistoric Exploitation of the Manatee in the Maya and Circum-Caribbean Areas. *World Archaeology*, *16*, (3), 337-353. Available at [http://www.jstor.org/discover/10.2307/124540?uid=3737808&uid=2&uid=4&sid=21103571312193] Downloaded 5 April 2013 from the JSTOR database.

Morales, D., Torres, L., Córdova, A. & Hernández, I. (2011). Catálogo de fauna de sitios arqueológicos aborígenes de Cuba. Instituto Cubano de Antropología (2011). Informe Final de Proyecto. Archivos del ICAN. (Unpublished).

Newsom, L. A. & Wing, E. S. (2004). *On land and sea: Native American uses of biological resources in the West Indies*. Tuscaloosa, USA: The University of Alabama Press.

Oviedo, G. (1851). *Historia General y Natural de las Indias, Islas y Tierra-Firme del Mar Océano*. Madrid. Imprenta de la Real Academia de la Historia.

Pérez, L. & Guarch, E. (2003). El medio ambiente precolombino en Cayo Bariay. Una perspectiva Arqueológica. *Revista Electrónica Ciencias Holguín, Año IX, No. 2*, Holguín. Available at [www.cienciasholguin.cu].

Pérez, L., Jardines, J. E. & Rodríguez, C. A. (2004). Estudio de los restos faunísticos en Los Buchillones. Economía y Medio Ambiente. *Archeofauna*, Tropical *Zooarchaeology*, *13*, 71-84.

Pérez, L. J., Guarch y col. (2012). Valoración de la Biodiversidad de la Región Centro Oriental de Cuba desde una perspectiva arqueológica. Departamento de Arqueología. CISAT, CITMA, Holguín. November 2012. (Unpublished).

Plazas, C. & Falchetti de Saenz, A. M. (1981). *Asentamientos Prehispánicos en el Bajo Río San Jorge*. Bogotá, Colombia: Fundación de Investigaciones Arqueológicas Nacionales.

Ramos, E. (2008). Fauna Arqueológica del Caribe Colombiano. Final Research Report, presented to the Fundación de Investigaciones Arqueológicas Nacionales. Bogotá. (Unpublished).

Ramos, E. (2010a). Arqueofauna y adaptacion humana en el Caribe Colombiano. In *Proceedings of the XXII Conference of the International Association for Caribbean Archaeology (IACA) – The Diversity of Caribbean Archeology* (pp. 627–645). Kingston, Jamaica: National Heritage Trust.

Ramos, E. (2010b). Los reptiles en la arqueología del Caribe Colombiano: Balance y perspectivas de investigación., In M. A. Gutiérrez, M. De Nigris, P. M. Fernández, M. Giardina, A. Gil, A. Izeta, G. Neme, & H. D. Yacobaccio (Eds.). *Zooarqueología a principios del siglo XXI: Aportes teóricos, metodológicos y casos de estudio* (pp. 583–592). Buenos Aires: Ediciones del Espinillo.

Ramos, E. & Archila, S. (2008). *Arqueología y subsistencia en Tubará. Siglos IX–XVI DC*. Bogotá, Colombia: Legis SA.

Ramos, E. (2009). *Más allá de la forma y la función: artefactos de hueso prehispánicos en Colombia*. Bogotá, Colombia: Fundación de Investigaciones Arqueológicas Nacionales: Universidad de los Andes – CESO.

Reichel-Dolmatoff, G. (1985). *Monsú. Un sitio arqueológico*. Bogotá, Colombia: Biblioteca Banco Popular.

Rivero, M. (1981). Pendientes aborígenes cubanos. *Revista de la Biblioteca Nacional José Martí*, *23*(1), 49-60.

Rives, A. & Febles, J. (1988). *Aproximación a una metódica interpretativa a los ajuares de sílex de las comunidades aborígenes de Cuba. Anuario de Arqueología*. Editorial Academia, Havana.

Rodríguez, R. & Vento E. (1989). *Algunos desdentados extinguidos de Cuba*. Havana, Cuba: Editorial Academia.

Rouse, I. (1942). *Archaeology of Maniabón Hills. Cuba*. New Haven, USA: Yale University Press.

Saldarriaga, G. 2006. Consumo de carnes en zonas cálidas de Nuevo Reino de Granada: cualidades cambiantes siglos XVI y XVII. *Fronteras de la Historia, No. 11*, 21-56.

Tabío, E. (1984). Nueva Periodización para el Estudio de las Comunidades Aborígenes de Cuba. *Islas*, *78*, 35 – 51.

Torres, L., Gutiérrez, D., González, J., Fernández, R. & Tavares, C. (2011). Reflexiones sobre el posible papel de las Ballenas en los modos de vida precolombinos del Caribe Insular, República Dominicana, un caso de ejemplo. *Cuba Arqueológica, 10.* Available at [http://www.cubaarqueologica.org/ index.php?q=node/55].

Ulloa J. & Corbea, J. (2011). *José Juan Arrom y la búsqueda de nuestras raíces.* Santiago de Cuba, Cuba: Editorial Oriente.

Valcárcel, R. (1999). Banes Precolombino: Jerarquía y Sociedad. *El Caribe Arqueológico, 3*, 84-94.

Valcárcel, R., Rodríguez, C., Pérez, L. & Guarch, J. J. (2003). Banes Precolombino. Catálogo de Objetos de uso ceremonial y de adorno corporal. Departamento de Arqueología CISAT. Holguín (Unpublished).

Vento Franco, G. (2001). La verdad sobre la Foca de Las Antillas. *Revista de Arqueología y Espeleología.* Available at [http://www.italia-cuba.speleo.it/1861_1-4.pdf]. No 1 – 4.

Villar, B., Gallo, J. P. & LeBoeuf, B. (1986). La foca Monje, Monachus tropicalis (Mammalis, Pinnipedia) definitivamente extinguida en México. *An. Inst. Biol. Univ. Nal. Autón. Mex Ser. Zool.*, *56*(2), 573-588.

Volkmer de Castilho, P. & Simões-Lopes, P. C. (2008). Sea mammals in archaeological sites on the southern coast of Brazil. *Revista do Museu de Arqueologia e Etnologia, São Paulo, 18*, 101-113.

In: Neotropical and Caribbean Aquatic Mammals ISBN: 978-1-63321-306-7
Editors: Muñoz, Götz and Ramos Roca © 2014 Nova Science Publishers, Inc.

Chapter 5

Human Use of Aquatic Mammals in Northern South America (Colombia and Panama): Some Evidence from Ethnographic and Ethnohistorical Records

Diana Rocío Carvajal Contreras*

Universidad Externado de Colombia, Facultad de Estudios de Patrimonio Cultural, Programa de Arqueología, Colombia

Abstract

Archaeology, ethnography and ethnohistory have been used as sources to understand the life of past Pre-Columbian people in Northern South America (Colombia and Panama). This chapter summarizes the use of aquatic mammals by human populations in Northern South America (Colombia and Panama) and is based on ethnographic and ethnohistoric accounts from the 16th to 19th century. Drawing on ethnohistoric sources, I first present the availability of and access to aquatic mammals for human populations. Secondly, ethnographic and ethnohistoric accounts are referred to in order to illustrate how people hunted and consumed meat and fat from these aquatic mammals. Thirdly, written records are drawn upon to illustrate the use of aquatic mammals in the manufacture

* E-mail: diana.carvajal@uexternado.edu.co

of tools and adornments. Finally, the ethnographic record from the 20th century is revisited to see if cultural attitudes toward these animals could have biased their consumption.

Keywords: Panama, Colombia, ethnohistory, ethnographic, aquatic mammals

INTRODUCTION

Archaeology, ethnography and ethnohistory have been used as sources to understand the life of past pre-Columbian people in Northern South America (Colombia and Panama).

The purpose of this chapter is to summarize the use of aquatic mammals by human populations in Northern South America (Colombia and Panama), from ethnographic and ethnohistoric accounts from the 16th to 20th century.

As is evidenced in this volume (see Ramos), prehistoric and historic human populations used aquatic mammals. Faunal remains deposited in archaeological sites show the use of these mammals for food and raw material to produce tools and adornments. However, not all aspects of the use of aquatic mammals are accessible through the archaeological records. Thus, in this text written sources have been used to complement information on past populations.

There is a great variety of written records related to information on aquatic mammals available through studies made by European chroniclers, travellers, and ethnographers (Cieza de León 1709; Colón 1892; de Carvajal 1892; de Castellanos 2004; Fernández de Oviedo 1535; Gumilla 1944; López de Gomara 1554; Simón 1892, Tovar 1994, 1995). However, the information in regards to the Amerindian use or consumption of aquatic mammals by pre-Columbian communities is scarce.

In archaeology, ethnohistoric information has been used as a source of analogies to interpret the archaeological record and to generate a rough knowledge of past cultures and their adaptations to their environment. However, Curet (2006) points out the dangers of relying only on these documents for the interpretation of past human cultures.

Firstly, in this chapter I will present the availability of and access to aquatic mammals for prehistoric and historic people using ethnohistoric sources. In the second section, ethnographic and ethnohistory accounts are drawn upon to illustrate how people hunted and consumed meat and fat from these aquatic mammals. Thirdly, written records are referred to with the aim of illustrating how aquatic mammals were used in the manufacturing of tools and adornments. Finally, the ethnographic record is revisited to see if cultural attitudes toward these animals could have biased their consumption.

About Availability of and Access to Aquatic Mammals for People in Northern South America (Colombia and Panama)

The obtainability of aquatic mammals for further exploitation as food or raw material by indigenous people involved on their part a deep ecological knowledge, not only of the habitat where each species could be found but the ethology of these animals. With this knowledge in mind, specific methods of capturing were applied depending on the species in question. (Correa et al. 2012; Patiño 1990). For the Europeans and Amerindians, this knowledge of the species and environment was also used for navigational purposes. Cristobal Colon (1892) mentions how a whale sighting was used as a sign to tell if the mainland was close, as was the case on his journey near Fernandina Island.

Free-swimming sperm whales have been recorded sporadically in regional waters as well as other aquatic mammals (Rasmussen et al. 2004: Table 7, Trujillo et al. 2013). 20th century records include Panama as a whaling ground for sperm whales (NOAA 2006), which are known to be susceptible to mass strandings, e.g., in the Gulf of California and Ecuador (Félix 2005; Vidal & Findlay 1986). Current biological information related to types of aquatic mammals, their locations, and geographical distribution is summarized in the following table:

Table 1. Species of aquatic mammals registered in the areav (Trujillo et al. 2013)

	Scientific name	Common name	Region	Remarks
Cetacea	**(SubordenMysticeti) Family Balaenopteridae**			
	Balaenoptera acutorostrata	Minke whale	Caribbean and Pacific	
	Balaenoptera borealis	Sei whale	Caribbean and Pacific	
	Balaenopteraedeni	Bryde´s whale	Caribbean and Pacific	

Table 1. (Continued)

	Scientific name	Common name	Region	Remarks
	Balaenoptera musculus	Blue whale	Caribbean and Pacific	
	Balaenoptera physalus	Fin Whale	Caribbean and Pacific	
	Megaptera novaeangliae	Humpback whale	Caribbean and Pacific	
	(SubordenOdontoceti) Family Physeteridea			
	Physeter macrocephalus	Sperm whale	Caribbean and Pacific	
	Kogia breviceps	Pigmy sperm whale	Caribbean	
	Kogia imus	Dwarf sperm whale	Caribbean and Pacific	
	Family Ziphiidae			
	Mesoplodon densirostris	Blainville's beaked whale	Caribbean and Pacific	
	Mesoplodon peruvianus	Pigmy beaked whale	Caribbean	
	Mesoplodon europeus	Gervais beaked whale	Caribbean	
	Ziphius cavirostris	Cuvier beaked whale	Caribbean and Pacific	
	FamiyDelphinidae			
	Sotalia fluviatilis	Bufeogris, Tucuxi	Amazon river	
	Sotalia guianenis	Marine Tucuxi	Caribbean	
	Steno bredanensis	Rough-toothed dolphin	Caribbean and Pacific	
	Peponocephala electra	Melon-headed whale	Caribbean and Pacific	
	Feres aattenuata	Pygmy killer whale	Caribbean and Pacific	
	Pseudorca crassidens	False killer whale	Caribbean and Pacific	
	Orcinus orca	Killer whale	Caribbean and Pacific	
	Globicephala macrorhynchus	Short-finned pilot whale	Caribbean and Pacific	

	Scientific name	Common name	Region	Remarks
	Grampus griseus	Risso's dolphin	Caribbean and Pacific Seas	
	Tursiopstruncatus	Bottlenose dolphin	Caribbean and Pacific	
	Delphinus delphis	Short-beaked common dolphin	Pacific	
	Delphinus capensis	Long-beaked common dolphin	Caribbean	
	Stenella attenuate	Pantropical spotted dolphin	Caribbean and Pacific	
	Stenella clymene	Clymene dolphin	Caribbean	
	Stenella coeruleoalba	Striped dolphin	Caribbean and Pacific	
	Stenella frontalis	Atlantic spotted dolphin	Caribbean	
	Stenella longirostris	Spinner dolphin	Caribbean and Pacific	
	Lagenodelphishosei	Fraser's Dolphin	Caribbean and Pacific	
	Family Iniidae			
	Inia geoffrensis	Bufeo, tonina, Amazon river dolphin	Amazon and Orinoco rivers	
	Sirenia			
	Trichechusinunguis	Amazonian manatee	Amazon river	
	Trichechusmanatus	West Indian manatee	Amazon and Orinoco rivers. Caribbean and Pacific.	

Table 1. (Continued)

Carnivora	Family Mustelidae			
	Scientific name	**Common name**	**Region**	**Remarks**
	Lontra longicaudis	Neotropical otter	Andes, Amazon and Orinoco rivers. Caribbean and Pacific	
	Pteronura brasiliensis	Giant otter	Amazon and Orinoco rivers	
	Family Otariidae			
	Arctocephalus australis	South American fur seal	Pacific	Occasional
	Arctocephalus galapagoensis	Galápagos fur seal	Pacific	Occasional
	Otaria flavescens	South American sea lion	Pacific	Occasional
	Zalophus californianus	California sea lion	Pacific	Occasional
	Family Phociidae			
	Monachus tropicalis	Caribbean monk seal	Caribbean	Extinct

Fernandez de Oviedo (1535) describes how Spanish sailors observed whales in the Pacific Ocean and compares these experiences with those of sailors in other parts of the world. Whale or dolphin sightings were reported near the Gulf of Orotiña, now the Gulf of Nicoya, between Nicaragua and Costa Rica, and near the Pacific coast of Panama, spotted as a result of their leaps in and out of the sea. Sperm whales were also observed on the Caribbean Sea between Trinidad Island and the main coast on the Gulf of Paria (De Armas 1888).

> "Déstas muchas veces las he hallado e visto en estas mares del Norte, entre aquestas islas e Tierra firme; e también en las costas de Tierra Firme tiene de la banda del sur, como más particularmente lo escribiré

cuando de aquella tierra, en la segunda parte d esta Natural Historia, proseguiré" (Fernandez de Oviedo 1535).1

De las Casas (1875:237) describes whale sightings during the expedition of Vasco Núñez de Balboa in the Pacific near the Gulf of Panama, Darien province. He also describes an accident near Fuerte Island on the Gulf of Uraba in Colombia. The big tail of a whale destroyed one of the boats of Francisco Pizarro:

> "Hiciéronse á la vela, seis meses después que allí habian entrado; salidos del golfo de Urabá, y siendo, cerca de la isla Fuerte, obra de 20 leguas, salidos á la mar, dio un golpe de mar al bergantín de Valenzuela, que lo metió con todos los que llevaba debajo del agua, donde, á vista de Pizarro y de los que con él iban oyendo los gritos dedos, todos se ahogaron; dijeron los del otro bergantín, que vieron una ballena ó otro pece muy grande, que con la cola les hizo pedazos el timón ógobernario."(De las Casas 1875:312)

There are rare mentions in the chronicles of sea lions or seals in the area. Fernandez de Oviedo (1535) for example, mentions that seals were seen near the Antilles and the mainland on the Caribbean coast (De Armas 1888). Oviedo makes a comparison between seals, sea lions and manatees, noticing the physical presence of breasts in all three species. Hermes Tovar (1994:429-430) mentions in his description of the governorate of Cartagena in 1571 that seals or sea lions are present on the Caribbean coast in Colombia, sharing their habitat of rivers and marshes with catfish, manatees, crocodiles and big turtles.

> "Lagunas y ciénagas y ríos, bagres y sabogas y manatíes en todos los ríos que salen a pacer la tierra, y su pescado es como carne de puerco, tiene el rostro como ternera de dos vacas y más de largo y tiene narizes por donde rebufan con dos ante puertas que se les tapa de manera que no les oça de entrar gota de agua... En las ciénagas ay lagartos grandes, caimanes y lobos marinos y tortugas grandes, los hijos de los lagartos los crían y los indios y los comen".

---

[1]" These (whales) have often been found and seen in the northern seas, between these Islands and the Mainland (Caribbean sea and the Antilles) ; and also on the shores of the mainland at the South band (Pacific ocean), I will in particular write about this land, and in the second part of this Natural history I will continue".

For some Europeans, in Brazil near Pernambuco, seals and other otariids and focids from the waist up were seen as similar to human beings and were even compared to monks, as was the case in the Antilles (De Armas 1888, Lopez de Gomara1554).

There are few mentions of dolphins by European chroniclers and most of these records refer to them as *tonina* and consider them to be fish not mammals (Colón 1832; De Armas 1888; Oviedo 1535).

The mention of manatees in the written record is more ample than other aquatic mammals (Cieza de León 1709; Castellanos; Fernández de Oviedo 1535; Gumilla 1944; López de Gomara 1554; Simón 1892, Tovar 1994, 1995). A description by Cristóbal Colón (1892:144) refers to manatees as mermaids, in Rio de Oro near Monte Cristi in the Antilles.

> "El día pasado cuando el Almirante iba al Río del Oro, dijo que vido tres serenas que salieron bien alto de lamar, pero no eran tan hermosas como las pintan, que en alguna manera tenían forma de hombre en la cara".

Certainly, Europeans would not have seen a creature like this before they arrived to the West Indies, as mentioned by D'Anghera (1912).

Figure 1. A manatee depiction (Ashton 1890:213).

D´Anghera added to these mythological interpretations by confusing manatees with Nereids or Tritons (Figure 1).

In the following excerpt some of the manatee´s physical characteristics are described, as well as its ¨domestic" or gentile behaviour:

> "A certain cacique of the region, Caramatexius by name, was very fond of fishing. Upon one occasion a Young fish of the gigantic species called by the natives "manati" was caught in his nets. I think this species of monster is unknown in our seas. It is shaped like a turtle and has four feet, but is covered with scales instead of shell. Its skin is so tough that it fears nothing from arrows, for it is protected by a thousand points. This amphibious creature has a smooth back, a head resembling that of a bull, and is tame rather than fierce. Like the elephant or the dolphin, it likes the companionship of men and is very intelligent. The cacique fed this young fish for several days with yucca bread, millet, and the roots the natives eat. While it was still young, he put it in a lake near to his house, as in a fish-pond. This lake, which had been called Guaurabo, was henceforth called Manati. For twenty-five years this fish lived at liberty in the waters of the lake, and grew to an extraordinary size. All that has been told about the lake of Baiae or the dolphins of Arion is not to be compared with the stories of this fish. They gave it the name of Matu, meaning generous or noble, and whenever one of the king's attendants, specially known by him, called from the bank Matu. Matu, the fish, remembering favours received, raised its head and came towards the shore to eat from the man's hand" (D'Anghera 1912:304).

Similar descriptions of the manatee´s remarkable physical characteristics, which are similar to human beings, specifically in relation to gender traits, are found in Colombian documents from near the Cesar, Orinoco and Magdalena rivers and Caribbean coast:

> "En las ciénagas que se hazen a los lados del arcabuco y Ribera del Río [Magdalena] que las ay de a dos, tres y quatro leguas se crían unos animales // que llaman manatís cuya cabeza es como un ternero, tienen unos braços como aletas de una tersia de largo y una quarta de ancho y unas como señales de uña de hombre, debajo de estos brazos asoma una punta de pecho como de mujer, la barriga como de dos bueyes, la cola en forma de abentador muy grande, las partes de generación como de hombre o mujer, andan en el agua y los ijuelos los traen debajo de aquellos brasetes o alillas, paçen yerba y salen a tierra a juntarse para la generación, no tienen más pie ni braço que aquellas aletas. Desde los hombros, pecho y barriga es muy ancho y de allí por

adelante ba engostando hasta la cola que se vuelve a ensanchar en forma de aventadero" (Tovar 1995:339).

Even though Europeans mentioned some characteristics (breast, hair etc.) which compared manatees to mammals such as cows or bulls, these amphibious creatures with weird extremities, dolphins and whales, were associated in the European folk taxonomy with fish (Cieza de León 1709; D'Anghera 1912; De las Casas 1875; De Carvajal 1542; De Castellanos 2004; Exquemelin 1969; Fernández de Oviedo 1535; Gumilla 1944; López de Gomara 1554; Simón 1892; Tovar 1994, 1995). These Europeans also documented life cycles, habitats and nutrition. Here are some examples:

"Hay en esta mar, en especial por estas islas, á la boca de los rios, entre el agua salada y dulce, los que llamaban los indios manatíes, la penúltima sílaba luenga: estos se mantienen de yerba, la que nace en el agua dulce junto á las riberas. Son tan grandes como grandes terneras, sin pies sino con sus aletas, con que nadan, y bien tienen tanto y medio como una ternera; no es pescado de escama, sino de cuero como el de las toninas ó atunes, ó como de ballenas; el que lo comiese delante quien no supiese qué era, en Viernes Santo, creería que comía propia carne, porque así lo parece; es muy más sabroso y precioso que ternera, mayormente los que se toman pequeños, echados en adobo como se suele comer la ternera." (De las Casas 1875: 304).

On the other hand, Gumilla (1944) describes the manatee from the Orinoco River, noticing its similarities with counterparts on the Caribbean coast and other mammals; however, he also classified it as a type of fish. He points out that stealth should be utilized while hunting the manatee due to its acute sense of hearing, something that had also been noticed by indigenous hunters:

"Es la figura del manatí, o vaca marina, muy irregular, y diversa de todo otro pescado. Ya dije, que se mantiene de la yerba, y ramas que se crían a las márgenes del río. La dentadura toda, y modo de rumiar, es propia de buey. También son muy semejantes a los del buey, su boca, y labios, con semejantes pelos a los que tiene también el buey junto a la boca. En lo restante de la cabeza no se le parece, porque los ojos son muy pequeños, y desproporcionados a su grande mole: sus oídos apenas se pueden distinguir con la vista; pero oye de muy lejos el golpe del remo:

por lo cual los pescadores bogan sin sacar el remo del agua, por no hacer ruido" (Gumilla 1944: 297)[2].

There is also mention of a dead manatee in the Amazon River during Francisco de Orellana´s expedition. Note in this quote the great amount of meat that a manatee could supply, which provided food for the group for six days:

> "Y fue así que un día, sobre tarde, pareció que venía por el río una vaca muerta tamaña como una mula, y, visto por el capitán, mandó a nuestros compañeros que se la trajesen y tomasen una canoa para traerla, y la trujeron, y se repartió por todos los compañeros de manera que a cada uno le cupo de comer para 5 ó 6 días, que no fue poco sino mucho remedio para todos. Esta vaca venía recién muerta, porque estaba caliente y no traía ninguna herida "(De Carvajal 1542:66).

There is no mention of other aquatic mammals, such as the species of the Mustelidae family, by the European chroniclers.

Human Consumption of Meat and Fat from Aquatic Mammals

Although there are several archaeological references to the exploitation of aquatic mammals for food in the area (Alvarez & Maldonado 2009; Angulo 1987; Arevalo & Maldonado 1990; Cooke & Jiménez 2009; García 1997; Legros 1992; Martinez et al. 2009; Plazas & Falchetti 1981; Reichel-Dolmatoff 1985; Wake et al. 2013; Wing 1980), mention in the European chronicles of the consumption of meat and fat from aquatic mammals in the area by indigenous people is rare.

The references that are provided are limited to the consumption of manatee; while few references are found in relation to the consumption of whales, seals, sea lions, dolphins and otters in Colombia and Panama.

[2] "It is the figure of the manatee or sea cow, very irregular, and different from all other fish. I have already mentioned that it feeds of the grass and branches that grow on the banks of the river. The denture, and way of chewing, is typical of ox. They are very similar to the ox, in the mouth and lips, and with similar hairs in the mouth to that of the ox. In the rest of the head there are no similarities because the eyes of the manatee are very small and disproportionate to the head mass. The ears can barely be distinguished, but it can hear from far away the sound of the boat paddle: the reason for which, so as not to make noise, bogan fishermen use the paddles without lifting them from the water".

References, however, are copious in relation to other areas of the Americas. For example, whale hunting is mentioned in relation to Yucatan, where a cord and spear were used and there are important mentions with regard to the consumption of otariids and phocids in the southern part of South America (De Armas 1888; Patiño 1990). There is no information on stranded animals or mustelids exploitation in Panama or Colombia (Patiño 1990).

Fernandez de Oviedo (1535) describes how the hunting of manatees was carried out. Pre-Columbian fishermen on the West Indies used a reversus or sucking fish and launched it from a canoe towards the manatee, using this fish as a kind of harpoon (Figure 2)[3].

De Armas (1888) does not believe that the hunting of the manatee described by Fernández de Oviedo in the Antilles is a true description. McKillop (1985), using ethnographic, ethnohistoric and archaeological evidences, suggests the hunting of the manatee was carried out with actual harpoons launched from canoes, as illustrated in the following quote by Fray Pedro Simón. This is similar to the method used by pre-Columbian inhabitants on the Colombian Caribbean coast and the marshes along the Magdalena River (Tovar 1995:440):

[3] "When the Indians wish to tame and keep any of these Reversus fish for their use in fishing, they catch it small and keep it always in salt water from the sea, and there they give it food and make it tame, until it is of the size which I have said or a little more, and fit for their fishing. Then they take it out to sea in the canoe or boat, and keep it there in salt water and fasten to it a delicate but strong cord. Then when there is seen a tortoise or any of the large fish which abound in these seas, or some of these Manatee or whatever it may be that happens to go on the surface of the water in such a way as to attract attention, the Indian takes this Reversus fish in his hand and strokes it with the other, and tells it to be manicato, which means strong and of good courage and to be diligent, and other words exhorting it to bravery, and to see to it that it dare to grapple with the largest and best fish that it may find there [where the fishing is to take place]. And when the Indian sees that the best time has arrived, he lets it go and even throws it in the direction of the large fish. Then the Reversus goes like an arrow and fastens itself on the side of a turtle, or on the belly, or wherever it can, and thus clings to it or to some other large fish. This one, when it feels itself seized by the little Reversus, flees through the sea in one direction or another; and in the meantime the Indian fisher-man lengthens the cord to its full length, which is many fathoms, and at the end of this is fastened a stick or cork that it may be for a signal or buoy which will remain on top of the water. In a little while the Manatee or turtle, to which the Reversus has attached itself comes to the shore, and then the Indian fisherman begins to draw the cord into his canoe or boat and when there are but a few fathoms left, he commences to draw it in carefully and slowly, guiding the Reversus and the prisoner to which it is attached until they reach land and the waves of the sea throw them out. The Indians who are engaged in the fishing leap out on land and if the prisoner is a tortoise they turn it over even before it has touched the" ground and place it high and dry because they are great swimmers; and if it is a manatee they harpoon, wound and kill it.

Figure 2. The mode of fishing and sailing by Pre-Columbian people (Benzoni 1857: 243).

> "Poséanlos con arpones desde las cancillas ó curianas, dándoles sirga hasta que se desangran. Cuando se ven perseguidos, arrojan con la cola tal espadaña de agua, que echan á pique una canoa" (Simón 1892: 296).

Several ethnographic references from the Handbook of South American Indians in the Circum-Caribbean and Tropical Forest tribe areas confirm harpooning as a method for hunting marine or riverine manatee by Arawaks, Achagua, Guamontey, Guamo, Panoans, Otomaco; Omagua, Cocama and Tucano (Morey 1975; Steward 1948a,b). Other groups had other hunting techniques, for example the Yaruro people caught manatee with hooks made of strong wood or iron (Steward 1948a:457).

On the Orinoco River the manatee was harpooned from a canoe paddled by the hunter's wife, while on the Amazon it was caught in a net and killed by driving a wooden plug up its nostrils (de Carvajal 1542; Gumilla 1944; Morey 1975; Steward 1948b:11). Another method was one used by the Chebero people and other groups on the Orinoco River. This method consisted of stretching a strong net across the opening of a weir. The hunter drove the animal to the opening, where it was caught in the net or speared by a man standing on a platform (Gumilla 1944; Patiño 1990; Steward 1948b:609). Amerindian people used manatee fat to cook their food (Steward 1948a, b). There is a brief mention of the exploitation of otters in the Guyanas, usually hunted with a blowgun or arrows from the canoes (Steward 1948b:827).

Although there are few references to Native American hunting methods of aquatic mammals for meat and fat, there are several references to European exploitation of manatees for these materials (Exequemelin 1969; Fernandez de Oviedo 1535; Tovar 1994, 1995*)*.

Fernandez de Oviedo describes how Spanish sailors hunted sea lions on land at night in the West Indies for their meat and used the fat as oil for cooking or to fill lamps:

> "… tiene una grosura, todo él en torno, tan ancha como una mano o altor de cinco dedos, todo rodeado, e a par del cuero, de una gordura de que se saca aceite muy bueno para arder en los candiles, e para guisar huevos e otras cosas, sin ningún rancio ni mal sabor" (Fernández de Oviedo 1535).

Numerous references to European hunting of manatees exist in the written record and the methods used usually replicated Native American techniques for the capture of manatee. Buccaneers described manatee meat consumption near Boca del Drago (Bocas del Toro) in Panama:

> "These fishes have very acute sense of hearing, so in taking them, the fishermen make not the least noise, nor row, unless very slightly. For this reason, they use certain instruments for rowing, called pagaros by the indians, by the Spaniards, canoa with which they row without any noise to frighten the fish. While they are fishing, they speak not one to another, only make signs. They dart with a javelin as they do tortoises; but the point of javelin somewhat differs, having two hooks at the end, and being longer. These fishes are from twenty to twenty-four long. Their flesh is

good to eat, being like in color to that of a land cow, but in taste, to pork. It has much fat, or grease, which the pirates melt, and keep in earihern pots to use instead of oil" (Exquemelin 1969:3093).

In Colombia along the Cesar River, chroniclers described manatee characteristics and their habitats. They also mention that the Tupes, the Caribe Indians and the Malibues consumed manatee meat and Spaniards used manatee fat for cooking or to fill lamps:

"...en el dicho rrío grande tres leguas atrás haze tres lagunas muy grandes que la primera se llama la Puxangua, será de más de veynte y cinco leguas de bojo, la una de Pancoche mayor que la laguna de Puxangua y la otra Tamalaguataca que tiene más de quarenta leguas de boxo y que en las dichas lagunas ay muy gran cantidad de pescado grueso e menudo y ay un género de pescado que tiene hocico como ternera e la cabeça e tiene los ojos pequeñitos sin orejas de grandor de quinze pies de largo es como un buye grande e la cola como una grande adarga llana e no tiene más de dos braçuelos cortos de un palmo e los dedos como de pato y es de color pardo y los indios) los matan para comer e los españoles se aprovechan de la manteca o azeite para comer y alumbrar en el candil" (Tovar 1994: 251).

Finally, European exploitation of aquatic mammals is characterized by a resulting long-term degradation of ecosystems, especially in the Caribbean (Jackson 1997; McClenachana et al. 2010). After the 17th century the manatee´s fat is still used to fill lamps and its hide used to make belts (Fals Borda 2002:16A). West Indian manatees were abundant not only in the Magdalena River´s marshes but also in the Atrato River. This mammal was hunted close to its extinction in order to feed slaves at Choco´s quarries and sailors or *bogas* of the Magdalena River. Preserved manatee meat was also commercialized for exportation until the 19th century (Baptiste–Ballera et al. 2002).

The Manufacture of Tools and Adornments from Aquatic Mammals

The history of the use of the remains of aquatic mammals (bone and hide) for tools and adornment has been mentioned in archaeological records from northern South America and its vicinities (Cooke 2005; Cooke and Jiménez 2010; Ladd 1964; Lothrop 1937; McKillop 1985; Ramos 2009; Wake et al. 2013; Wing 1980).

Early evidence of modified manatee bones was found in a pre-Columbian midden at Cerro Mangote in Panama (Cooke 2005) around 8000-6000 years ago and in Puerto Chacho; Monsú and Guajaro in Colombia (Alvarez & Maldonado 2009; Angulo 1988; Arevalo & Maldonado 1990; Legros 1992) around 5,000 years ago.

Later, in three Panamanian archaeological sites, Sitio Conte, El Hatillo and El Caño elite cemeteries, carved manatee bones were identified (Cooke 2004a; Ladd 194; Lothrop 1937; Mayo & Mayo 2013). A monk seal tooth was found at Cerro Juan Diaz on the shore of Parita Bay in Panama (Cooke 2004a). At Sitio Conte, carved objects identified by Lothrop (1937) as whale tooth ivory from the lower jaw of a female sperm whale (*Physeter macrocephalus*) were found.

Historical accounts mention briefly the use of aquatic mammal remains as raw material for adornments and tools. Fernández de Oviedo (1535) mentions that Europeans used sea lion hides to manufacture belts and bags (De Armas 1888).

> "… digo que las cintas e correas que se hacen del cuero dél para ceñirse los hombres, o para bolsas, o para lo que quieren"

The following author also describes the use of the manatee´s hide for belts and shoes:

> "y el cuero es tan gordo como un dedo, e curándolo al sol, se hacen dél buenas correas e suelas para zapatos e para otros provechos. Y la cola dél, de la cintura que he dicho adelante, toda ella hácenla pedazos e tiénenla cuatro o cinco días más al sol (la cual paresce como nervio toda ella), e desque esta enjuta, quemanla en una sarten, o mejor diciendo, fríenla e sacan della mucha manteca…" (Fernandez de Oviedo 1535).

The manatee´s hide was used also by pre-Columbian inhabitants on the Caribbean coast and Orinoco River in the making of shields (de Carvajal 1542; Gumilla 1944):

> "Es pescado desnudo, de color pardo, pero tan fuerte su cuero, que suelen hacer de él los indios rodelas que no las pasa una flecha…" (Simón 1892:296).

Europeans reported the Pre-Columbian use of amber from sperm whales to produce artefacts in the Antilles, but there are no historical accounts of their use in the area (De Armas 1888).

Although, as is mentioned above, there is archaeological evidence of their use and it is presumed that amber was found on the beach by Indigenous people. There is no other evidence of the use of other aquatic mammals to produce artefacts.

In contrast, ethnographical accounts describe the use of aquatic mammals in the area for several other purposes. In the Handbook of South American Indians, manatee´s hides were used by the Juri, Pase, Guayupe and Sae people for defense.

They used round shields made of a single piece of manatee hide (Steward 1948a:389, Steward 1948b:709). Other Amerindian groups of the Tropical forest area, such as the Mura, Piraha; Canamar, Curina and Yurimagua celebrated a cult reminiscent of the "Yurupary feasts" of the Tucano and Arawakan tribes of the Cariary-Vaupes, using manatee hides for flagellation with a long leather thong (Steward 1948b).

Aquatic Mammals and Cosmogony

The relationship between Prehistoric and historic people and the hunting of mammals may have been influenced by cultural attitudes related to religion or cosmology (Ingold 1988).

The Spanish wrote that manatee meat tasted like cow or pork, and that people with syphilis should avoid it (Simon 1892, Tovar 1994, 1995). Here is an excerpt from a "relación" (a type of report) near Tamalameque (Colombia):

> "Desta cola se saca comúnmente una arroba de manteca que en todo es como el puerco, salvo que es más liquida. Sírvese della para alumbrar y para las comidas, aunque a los tocados de morbo gálico (sífilis) les hace notabilísimo daño y así huyen della" (Tovar 1994:293).

The Spanish also mentioned that the use of rings made of manatee bones were thought to cure haemorrhoids.

> "...son sus huesos medicinales, de que se hacen sortijas para el dedo anular contra la enfermedad de almorranas; en otras partes le llaman peje-buey, por la cabeza" (Simón 1892:296).

Gumilla (1944:297) mentions that manatees from the Orinoco River were not only appreciated for their meat and fat, as the bones from the skull were used to cure a blood illness.

> "...y el casco de la cabeza, tiene una chocozuela redonda, del tamaño de la bola de truco: este hueso es remedio experimentado contra flujos de sangre; y para este efecto se busca, y encarga con ansia".

Fernandez de Oviedo (1535) and Lopez de Gomara (1554) mention that these skull bones of the West Indies manatee were useful for curing gall bladder:

> "Estos animales tienen ciertas piedras o huesos en la cabeza entre los sesos o meollos. La cual piedra es mu útil para el mal de la hijada, según acá se platica e afirman personas tocados de está enfermedad" (Fernandez de Oviedo 1535).

Fernandez de Oviedo (1535) also mentions that belts made of sea lion hide were used to cure back pain. There are no other mentions of aquatic mammals and their use as medicine.

It is very difficult to assess the indigenous meaning assigned or cultural attitudes toward aquatic animals, since ethnohistoric records are not explicit. However, animal icons depicted in pottery, shell adornments and other media can give some ideas with regard to indigenous semiotic systems (Cooke 2004b).

Archaeological information in Panama indicates that aquatic mammals were used for food, tools and ornaments but were not used in pottery or shell adornments (Cooke 2004b; Helms 1995). In contrast, Anne Legast describes some archaeological material from Colombia and notes some biological morphological correlations between the pottery, metal work and the otter`s tail. As she (Legast 1993:87) suggests, a Ilama ceramic vessel has a multiple zoomorphic representation that includes a tail similar to a giant otter´s tail (*Pteronura brasiliensis*). Otter depictions are also observed in Muisca metalwork (Legast 1998:79) and otter remains were found at Sabana de Bogota (Correal 1981).

However, ethnographic accounts are copious with regards to aquatic mammals and their relation to indigenous and mestizo ways of thinking. For example, the bufeo or tumuli (*Sotalia fluviatilis*) is considered a good animal, while the tonina (*Inia geoffrensis*) is thought to evil or a mischievous. Both are considered taboo for Amazonian and mestizo people (Cravalho 1999; Reichel 1989).

It was thought that toninas kidnapped people, mainly women, and took them to the bottom of the river. Their teeth and fat were used in witchcraft. Dolphins were not hunted and their meat and fat and were not used for food. If eaten, it was thought that heartburn would result (Steward 1948b: 532). Its reddish colour was associated with European conquerors (Rodríguez 1981; Trujillo & Diaz Granados 2012). Ticuna Indians considered tonina are similar to a human being and to be the manatee´s great-grandchild and the Yukuna Matapí considered river dolphins to be as treacherous as human beings (Omacha 2004; Reichel 1989).

The Amazonian manatee (*T. inungis*) was thought by local populations to help in the treatment and cure of the following maladies, amongst others: sprains, asthma insect bites, menstrual cramps etc. The West Indian manatee (*T. manatus*) also contained medical properties which it was thought could cure arthrosis, luxation, and wounds, amongst others (Alves & Alves 2011). The Caribes believed that if indigestion had been caused by eating crabs, they could eat, as an antidote, the flesh of the manatee or tortoise, which besides relieving the ailment would also cause the body to break out in pimples (Steward 1948a: 563).

Some Amazonian groups such as the Mura and Conibo associated the manatee with the Southern Cross constellation, which was "swimming" in the river or Milky Way (Steward 1948b:265, 595). Tukano and Arawak people, in ceremonies similar to those of the Yurupari, belted young boys with a whip of manatee hide to make them strong (Steward 1948b: 704). Uitoto and Ticuna people called manatee river tapir (Chacon 2013; Marín & Becerra 2006).

Finally, Europeans humanized manatees because of their physical characteristics, as did mestizo people later in some parts of the Colombian Caribbean lowlands along the Magdalena River. Some of these sailors along the Magdalena River even had intercourse with manatees because of their resemblance to women:

> "Desembarcábamos con nuestra estera y el toldo y nos acostába- mos a descansar no lejos de una hoguera que prendíamos para alejar los mosquitos y los tigres. Algunos compañeros se enterraban en la arena para dormir mejor. Otros se alejaban a escondidas en búsqueda de alguna manatí grande para culear con ella, porque, como tú sabes, ese animal tiene chucha y tetas, y se sale del rio cada mes para pasar tres días de regla" (Fals Borda 1979:47a).

There are no other ethnographic accounts of other aquatic mammals but afro-descendent communities in the Caribbean Sea and Pacific Ocean humanized dolphins, humpback whales, and sperm whales. They were recognized by the sounds that they produced and were considered intelligent human beings. However, they were also considered treacherous and bold as well as calculating. They were not hunted or consumed (Correa et al., 2012).

CONCLUSION

This review of the available ethnohistoric and ethnographic record has demonstrated the diverse use of aquatic mammals by Amerindian and European populations from the 16th to 20th century. These records can be useful in drawing correlations between the information therein and the presence or absence of these mammals in the archaeological record. Not only can they help us to understand past life-ways but they can also help us to understand the present geographical distribution of some of these animals.

As it can be seen in table 1, in the Pacific Ocean and in the Caribbean Sea, as well as in rivers like the Orinoco, the Magdalena, the Atrato and the Amazon, we are presented with a great diversity and presence of aquatic mammals. As mentioned by Curet (2006), the chroniclers had a skewed vision of Amerindian reality and this bias is demonstrated in the absence of information relating to indigenous sightings. The historical records show that the presence of these animals was still distinguishable. Europeans considered some of them to be fish or mythological creatures and they often did not distinguish between species. Their descriptions can obscure the diversity of aquatic animals that was available to prehistoric people in the area as most of the references are related to manatee.

These same chronicles and ethnograpic references show that Amerindian populations had access to these aquatic resources for food or raw materials through hunting. Capture suggests that they were used only occasionally as sources of meat and fat and that these animal products were sourced from other animals. The Indians used bone, fat and meat from aquatic animals occasionally for food but more often for the initiation of Yurupari type ceremonies or as offerings of elite burials or to manufacture shields for war activities. Occasional stranded sperm whales or seals were the source of the ivory and teeth used for elite offerings at the Panamanian archaeological sites of Sitio Conte, Cerro Juan Díaz, El Hatillo and El Caño.

It was only with the arrival of European conquistadors that consumption became intensive, not only for food but also for raw materials for tools and ornaments.

This was until some of these animals became extinct or decimated (Baptiste-Ballera et al. 2002; Gotz & Sierra 2011; McClenachana et al. 2010).

The reason why pre-Columbian people occasionally used these aquatic animals in the area is found in zoo archaeological information. The pattern in Colombia and Panama is that the aquatic animals consumed by the Amerindian population came from the shallow waters of the coastline including the mangrove, beaches of the coast, or the forests near the intermediate zone between the coast and the inland region. There is no evidence that sailing was carried out for the purpose of fishing in deep waters, since none of the taxa found occurs exclusively in pelagic waters, although some species found in samples occasionally inhabit or travel through deep water. The white tailed deer (*Odocoileus virginianus*) was by far, the most consumed mammal in the area (Carvajal Contreras in press; Cooke & Martín 2010; Cooke & Sánchez; Peña 2010).

Marvin Harris (1989) argues that certain animals were avoided as sources of food, as the energy involved in their capture was not cost beneficial; a fact which was often expressed through certain cultural taboos. It might be feasible to think that these are the reasons for the scarcity of reference to aquatic mammals in the written record.

However, as mentioned above, the use of aquatic animals may have been related to cultural attitudes regarding religion or cosmology. Cultural regulations of preference for or avoidance of certain animal resources has been discussed in anthropology under different theoretical approaches (Meyer-Rochow 2009).

It might be the case that as indigenous populations expressed a cultural understanding of ecological factors and the importance of their balance, often influenced by social rules relating to gender, status etc, they caused less long-term damage to species through hunting (Arhem1990; Meyer-Rochow 2009). This was clearly not the case in relation to Europeans´ consumption of these mammals, which lead to the extinction of some and current geographical distribution of others.

To conclude, the present ethnohistoric and ethnographic information, although in many cases it is biased, may generate useful models to improve our interpretation and understanding of past life ways and the use of aquatic mammals.

Acknowledgment

First, I want to express my gratitude to the editors for allowing me to participate in this volume. Thanks to Hermes Tovar, Peter Stahl, Richard Cooke, Juan Vargas and José Luis Socarrás for their advice and information on bibliographic references.

References

Alvarez, R. & Maldonado, H. (2009). Arqueofauna Encontrada en Puerto Chacho, Sitio Arqueológico del Caribe Colombiano (3300 A. C.). *Revista de la Academia Colombiana de ciencias exactas, físicas y naturales, 33* (128), 407-424.

Alves, R. & Alves, H. (2011). The faunal drugstore: Animal-based remedies used in traditional medicines in Latin America. *Journal of Ethnobiology and Ethnomedicine,* (7)9,1-43.

Angulo, C. (1987). Guájaro, un modo de vida en la arqueología del norte de Colombia. *Anuario Científico Universidad del Norte,* VI, 93-105.

Angulo, C. (1988). *Guájaro en la arqueología del norte de Colombia.* Bogotá, Colombia: FIAN-Banco de la República.

Arévalo, H. & Maldonado, H. (1990). Una contribución al Formativo Temprano en Colombia. Tesis Profesional. Universidad Nacional de Colombia. Depto. de Antropología, 192 p.

Arhem, K. (1990). Ecosofia Makuna. In Correa, F. (Ed.)., *La Selva Humanizada* (pp. 105 -122). Bogotá: ICAN-Fondo Editorial CEREC.

Ashton, J. (1890). *Creatures in Zoology.* London, UK: Printed by John C. Nimmo 14, King William Street, Strand.

Baptiste- Ballera, L. Hernández, S. Polanco, R. & Quiceno M. (2002). La fauna silvestre colombiana: una historia económica y social de un proceso de marginalización. In Ulloa A. (Ed.). *Rostros culturales de la fauna: las relaciones entre los humanos y los animales en el contexto colombiano.* Bogotá, Colombia: ICANH Fundación Natura.

Benzoni, G. (1857). *History of the New World.* London, UK: The Hakluyt Society.

Carvajal-Contreras, D. (in press). Los moluscos y la arqueología. Análisis preliminar de tres sitios arqueológicos en el Canal del Dique Colombia. In

Boletin Cientifico CIOH (BCCIOH), Centro de Investigaciones Oceanográficas e Hidrograficas del Caribe (CIOH).

Chacon, T. (2013). On Proto-Languages and Archaeological Cultures: pre-history and material culture in the Tukanoan Family. *Revista Brasileira de lingüística Antropológica,* (5)1, 217-245.

Cieza de León, P. (1709). *The seventeen years travels of Peter de Cieza, through the mighty kingdom of Peru, and the large provinces of Cartagena and Popayan in South America: from the city of Panama, on the isthmus, to the frontiers of Chile*. Now first tr. from the Spanish, and illustrated with a map, and several cuts. London, UK: Edmund Poley of Badley.

Colón, C. (1892). *Relaciones y Cartas*. Biblioteca Clásica, Tomo CLXIV. Madrid, España: Librería de la Viuda de Hernando y C.

Cooke, R. (2004a). Rich, poor, shaman, child: animals, rank and status in the "Gran Coclé" culture area of pre- Columbian Panama. In O´Day, S., Van W. Neer & A.Ervynck, (Eds.). *Behaviour Behind Bones. The Zooarchaeology of Ritual, Religion, Status and Identity* (pp. 271-284). Oxford, UK: Oxbow books.

Cooke, R. (2004b). Observations on the religious content of the animal imagery of the "Gran Coclé semiotic tradition of pre-Columbian Panama. In O´Day, S., W. Van Neer, & A. Ervynck (Eds.). *Behaviour Behind Bones. The Zooarchaeology of Ritual, Religion, Status and Identity* (pp. 114-127). Oxford, UK: Oxbow books.

Cooke, R. (2005). Prehistory of Native Americans on the Central American Land Bridge: Colonization, Dispersal, and Divergence. *In Journal of Archaeological Research, 13(2),* 129-187.

Cooke, R. & Jiménez, M. (2009). Fishing a pre-Hispanic settlements on the Pearl Island archipelago (Panama, Pacific), 1: Pedro Gonzalez Island (4030-3630 cal BCE). In Makowiecki, D., S. Hamilton-Dyer, I., Riddler,, N. Trzaska-Nartowski & M. Makohonienko (Eds.). *Fishes-Culture-Environment Through Archaeoichthyology, Ethnography and History: The 15th Meeting of the ICAZ Fish Remains Working Group (FRWG)* (pp.167-171). Poznan & Turun, Poland: Bogucki Wydawnictwo Naukowe, (Srodowiskoi Kultura = Environment and Culture).

Cooke, R. & Jiménez, M. (2010). Animal-Derived artefacts at two Pre-Columbian sites in the Ancient Savannas. An Update of their Relevance to Studies of Social Hierarchy and Cultural Attitudes Towards Animals. In Campana, D., Crabtree, P., de France, S., Lev-Tov, J. & Choyke, A. (Eds.), *Anthropological Approaches to Zooarchaeology: Complexity,*

Colonialism and Animal Transformations (pp. 30-55). Oxford, England: Oxbowbooks.

Cooke, G. & Martín, J. (2010). Arqueozoología en la Baja América Central (Nicaragua, Costa Rica y Panamá). In G. Mengoni Goñalons, J. Arroyo Cabrales, Ó. Polaco & F. J. Aguilar (Eds.). *Estado actual de la arqueozoología latinoamericana/ Current advances in Latin-American archaeozoology* (pp. 105-131). Mexico City, Mexico: Instituto Nacional de Antropología e Historia. Consejo Nacional de Ciencia y Tecnología. International Council of Archaeozoology. Universidad de Buenos Aires.

Correa S., Turbay, S. & Vélez, M. (2012). Conocimiento ecológico local sobre ecosistemas marinos en dos comunidades costeras: El Valle y Sapzurro. *Revista Gestión y Ambiente,* 15(2), 17-32.

Correal, G. (1981). *Evidencias culturales y megafauna Pleistocénica en Colombia.* Bogotá, Colombia: FIAN- Banco de la República.

Cravalho, M. (1999). "Shameless creatures: an ethnozoology of the Amazon river dolphin". In *Ethnology,* (38)1, 47-58.

Curet, A. (2006). Las crónicas en la arqueología de Puerto Rico y del Caribe. *Caribbean Studies*, 34(1), 163-199.

D'Anghera, P. (1912). *De Orbe Novo The Eight Decades.* Translated from the Latin with Notes and Introduction By Francis Augustus MacNutt In Two Volumes. Volume One G. P. New York, USA & London, England: Putnam's Sons New York and London The Knickerboker Press.

De Armas, J. (1888). *La zoología de Colón y los primeros exploradores de América.* Habana, Cuba: Establecimiento Tipográfico O´Reilly N.9.

De Carvajal, G. (1542). *Descubrimiento del río de las Amazonas por el Capitán Francisco de Orellana,* Relación de Gaspar de Carvajal. Documento regalado a la Sección de Manuscritos de la Biblioteca Nacional de España por el Duque de T'Serclaes el 5 de mayo de 1961, Madrid.

De Castellanos, J. (2004). *Elegía de Varones Ilustres de Indias.* Edición y Prólogo de Luis Fernando Restrepo (Ed.). Bogotá, Colombia: Editorial Pontificia Universidad Javeriana.

De las Casas, B. (1875). *Historia de las Indias.* Madrid, España: Imprenta de Miguel Ginesta.

Exquemelin, O. (1969). *The buccaneers of America* [1678], Translated by Alexis Brown. Mineola, New York: Dover Publications, Inc.

Fals Borda, O. (1979). Historia doble de la Costa, tomo I: Mompox y Loba. Bogotá, Colombia: Carlos Valencia Editores.

Fals Borda, O. (1986). *Retorno a la madre tierra, e Historia Doble de la Costa T. 4*. Bogotá, Colombia: Carlos Valencia Editores.

Félix, F. (2005). La investigación de mamíferos marinos en la costa de Ecuador y perspectivas. In, Garzón, F., et al., (Eds.), *Ordenamiento Territorial Costero-Marino y Declaratoria de Areas Protegidas en Salinas* (pp. 1-17). Seminario-Consulta, Salinas, Ecuador, 15 de julio de 2005. Informe y Trabajos Presentados. Salinas, Ecuador.

Fernández de Oviedo & Valdés, G. (1535). *Historia general y natural de las Indias: Part 1* [online], December 2, 2013. Available from http://www.ems.kcl.ac.uk/content/etext/e026.html

García, M. (1997). Zoo-arqueología del Formativo Temprano de la costa Caribe: Un acercamiento a la dieta del grupo prehistórico de Puerto Chacho. *Tesis Profesional.* Universidad Nacional de Colombia. Depto. de Antropología, 145 p.

Gotz, C. & Sierra, T. (2011). La arqueofauna de Xcambó, Yucatán, México. *Antípoda, Revista de Antropología y Arqueología 13*, 119-145.

Gumilla, J. (1944). *El Orinoco Ilustrado: Historia Natural, Civil y Geográgica de este Gran río.* Biblioteca Popular de Cultura Colombiana. Tomo 1. Bogotá, Colombia: Editorial ABC.

Helms, M. (1995). *Creations of the Rainbow Serpent: Polychrome Ceramic Designs from Ancient Panama.* Albuquerque, USA: University of New Mexico Press.

Harris, M. (1989). *Bueno para comer Enigmas de alimentación y cultura.* Madrid, España: El libro de bolsillo Antropología Alianza Editorial.

Ingold, T. (Ed.) (1988). *What is it an animal? One World Archaeology.* London, UK: Unwin Hyman.

Jackson, J. (1997). Reefs since Columbus. *Coral Reefs, 16,* Suppl., S23- S32.

Ladd, J. (1964). *Archaeological investigations in the Parita and Santa María zones of Panama. Smithsonian Institution Bureau of the American Ethnological Bulletin 193*, Washington DC., USA: Smithsonian Institution.

Legast, A. (1993). *La fauna en el material precolombino Calima.* Bogotá, Colombia: Fundación de Investigaciones Arqueológicas Nacionales. Banco de la República.

Legast, A. (1998). La fauna Muisca y sus símbolos. *Boletín de Arqueología* 13 (3).

Legros, T. (1992). Puerto Chacho et les premiers céramistes americains – Nouvehes données sur le formatif ancien du litoral Caribe de Colombie.

These de Doctorat. Univ. de Paris 1, Panthéon - Sorbonne, Tomo 1: 367 p., Tomo II: 157 p.

López de Gomara, F. (1554). *La historia general de las Indias, con todos los descubrimientos, y cosas notables que han acaescido en ellas, dende que se ganaron hasta agora*. Madrid, España: Editorial Casa de Juan Stelsio.

Marín, P. & Becerra, E. (2006). Los habitantes del mundo sub-acuático. Ritual y mítica Murui Iye Fia Rafue. *Forma y Función,* 19, 103-124.

Martinez, M., Jiménez, M. & Cooke, R. (2009). Fishing at pre-Hispanic settlements on the Pearl Island archipelago (Panama, Pacific), II: Bayoneta Island (900-1300 CE). In: Makowiecki, D., S Hamilton-Dyer, I. Riddler, N. Trzaska-Nartowski & M. Makohonienko (Eds.). *Fishes-Culture-Environment Through Archaeoichthyology, Ethnography and History* (pp.172-175). Poznan: Poland: Bogucki Wynadwictwo Naukowe, (Srodowiskoi Kultura = Environment and Culture).

Mayo J. & Mayo, C. (2013). El descubrimiento de un cementerio de elite en el Caño: indicios de un patrón funerario en el valle de Río Grande, Coclé, Panamá. *Arqueología Iberoamericana,* 20, 3–27.

McClenachana L., Hardtb, M., Jackson, J. & Cooke, R. (2010). Comment on Julio Baisre's "Setting a Baseline for Caribbean Mounting Evidence for Historical Overfishing and Long-term Degradation of Caribbean Marine Ecosystems: Fisheries. *The Journal of Island and Coastal Archaeology,* 5,165-169.

McKillop, H. (1985). Prehistoric exploitation of the manatee in the Maya and circum-Caribbean areas. *World Archaeology,* 16(3), 337-353.

Meyer-Rochow, V. (2009). Food taboos: their origins and purposes. *Journal of Ethnobiology and Ethnomedicine*, 5:18 doi:10.1186/1746-4269-5-18.

Morey, N. (1975). Ethnohistory of the Colombian and Venezuelan Llano. PhD dissertation. University of Utah, Ann Harbor.

NOAA [National Oceanic and Atmospheric Administration]. (2006). Draft recovery plan for the sperm whale. Office of Protected Resources, National Marine Fisheries Service, Silver Spring MD. Available at www.nmfs. noaa.gov/pr/pdfs/recovery/draft).

Omacha [Fundación Omacha] (2004). *Historias de bufeos*. Bogotá, Colombia: Ediciones Antropos Ltda.

Patiño, V. (1990). Historia de la Cultura Material en la América Equinoccial, Tomo 5: Tecnología. [Online], December 13, 2013, Available from http://www.banrepcultural.org/blaavirtual/historia/equinoccial_5_recursos -industria/inicio.htm

Peña, G. (2010). Origen y desarrollo de la arqueozoología colombiana. In G. Mengoni Goñalons, J. Arroyo Cabrales, Ó. Polaco & F. J. Aguilar (Eds.). Estado actual de la arqueozoología latinoamericana/ Current advances in Latin-American archaeozoology (pp. 93-104). Mexico City, Mexico: Instituto Nacional de Antropología e Historia. Consejo Nacional de Ciencia y Tecnología. International Council of Archaeozoology. Universidad de Buenos Aires.

Plazas, C. & Falchetti, A. (1981). *Asentamientos prehispánicos en el bajo río San Jorge*. Bogotá, Colombia: FIAN- Banco de la República.

Ramos, E. (2009). *Más Allá de la Forma y Función. Artefactos de Hueso Prehispánicos en Colombia.* Bogotá, Colombia: FIAN- Banco de la República - Uniandes, CESO.

Rasmussen, K., Calombakidis, J. & Steiger, G. (2004). *Humpback whales and other marine mammals off Costa Rica and surrounding waters,* 1996-2003. Report of the oceanic Society Field Season in Cooperation with Elderhostel Volunteers, sponsored by Oceanic Society Expeditions, Available at www.oceanic-society.

Reichel- Dolmatoff, G. (1985). *Monsú: un sitio arqueológico*. Bogotá, Colombia: Biblioteca Banco Popular.

Reichel, E. (1989). La danta y el delfín: manejo ambiental e intercambio entre dueños de maloca y chamanes. El caso yucuna-matapí (Amazonas). *Revista de Antropología*, (5)1-2, 67-134.

Rodríguez, M. (1981). *Muestra de Literatura Oral en Leticia Amazonas T. LVII.* Bogotá, Colombia: Publicaciones del Instituto Caro y Cuervo.

Simón, P. (1892). *Noticias Historiales de las Conquistas de Tierra Firme en las Indias Occidentales*. Bogotá, Colombia: Casa Editorial de Medardo Rivas.

Steward, J. (Ed.) (1948a). *Handbook of South American Indians: The Circum-Caribbean Tribes*. Washington, DC., USA: Smithsonian Institution.

Steward, J. (1948b). *Handbook of South American Indians: The Tropical Forest Tribes*. Washington, DC., USA: Smithsonian Institution.

Tovar Pinzón, H. (1994). *Relación y Visitas a los Andes S. XVI. Región del Caribe T. 2*. Bogotá, Colombia: Colcultura, Instituto Colombiano de Cultura Hispánica.

Tovar Pinzón, H. (1995). *Relación y Visitas a los Andes S. XVI. Región Centro-Oriental T. 3*. Bogotá, Colombia: Colcultura, Instituto Colombiano de Cultura Hispánica.

Trujillo, F., Gärtner, A., Caicedo D., & Diazgranados, M. (Eds.) (2013). *Diagnóstico del estado de conocimiento y conservación de los mamíferos acuáticos en Colombia. Ministerio de Ambiente y Desarrollo Sostenible.* Bogotá, Colombia: Fundación Omacha, Conservación Internacional y WWF.

Trujillo, F. & Diaz Granados, M. (2012). *Delfines de Río: embajadores de la conservación en la Amazonía y la Orinoquía.* Bogotá, Colombia: Fundación Omacha.

Vidal, O., & Findley, L. (1986). Recent stranding of sperm whales in the Gulf of California, Mexico. *Journal of Mammalogy,* 67, 770-771.

Wake, T., Doughty, D. & Kay, M. (2013). Archaeological Investigations Provide Late Holocene Baseline Ecological Data for Bocas del Toro, Panama. *Bulletin of Marine Science,* 89(4), 1015–1035.

Wing, E. (1980). Aquatic fauna and reptiles from the Atlantic and Pacific sites. In Linares O. F., & A. J. Ranere (Eds), *Adaptive Radiations in Prehistoric Panama, Peabody Museum Monographs 5* (pp. 194-215). Cambridge, USA: Harvard University Press.

In: Neotropical and Caribbean Aquatic Mammals ISBN: 978-1-63321-306-7
Editors: Muñoz, Götz and Ramos Roca

Chapter 6

ZOOARCHAEOLOGICAL EVIDENCE OF OTARIIDS IN CONTINENTAL COAST OF PATAGONIA, ARGENTINA: OLD AND NEW PERSPECTIVES

Florencia Borella*
INCUAPA – CONICET, Facultad de Cs. Sociales – UNCPBA, Argentina

ABSTRACT

While there is frequent mention of the presence of sea lions in many coastal Patagonian Late Holocene assemblages, zooarchaeological analysis of these have been limited, mainly because of the lack of proper osteological studies. This is a common problem to archaeological marine mammal bone analysis across the world. In this chapter, we present the current state of pinniped zooarchaeological studies from the Atlantic coast of continental Patagonia, Argentina. To begin with, we highlight the manner in which pinnipeds have been registered traditionally in archaeology. Primarily, this entailed a simple identification of the presence of these animals in the zooarchaeological record, without then entering into a discussion on the nature of this evidence or further analysis towards exploring the role of pinnipeds in the subsistence strategies of past human populations.

* E-mail: fborella@soc.unicen.edu.ar

Since then, a new way of tackling pinniped remains has emerged. This new perspective seeks to develop varied analytical pathways towards resolving known problems in the taxonomical identification of pinnipeds. This is then used to answer specific research questions such as past pinniped exploitation strategies. This new perspective is made-up of a range of different analytical methods –osteometric studies, bone fusion chronology, identification of diagnostic traits – that are currently being developed. These are geared towards being able to identify more precisely within the wider category of pinnipeds. Another stated aim of this new approach is the possibility of determining by age classes and sex within each species. This will permit inferences on exploitation strategies and types of prey caught, thus providing interpretations on the role that these marine mammals had on continental Patagonian hunter-gatherer subsistence during the Late Holocene.

Introduction: First Mention of Pinnipeds in the Archaeological Record from Continental Patagonia

Archaeological research on the Atlantic coast of continental Patagonia frequently mentions the discovery of sea lions remains in sites that show evidence for exploitation of coastal resources[1]. These marine mammals were usually listed as 'pinnipeds', reflecting the fact that these investigations were non-specific and summarized different lines of archaeological evidence (e.g. Gómez Otero et al. 1999; Sanguinetti de Bórmida 1999; Castro et al. 2000; Gómez Otero 2007; Moreno 2008; amongst others). Aside from a few specialized zooarchaeological studies (Miotti 1998, Moreno et al. 1998), work on coastal faunal assemblages started later especially during this last decade. This is in marked difference with the trajectory of pinniped zooarchaeological studies in Southern and maritime Patagonia, where investigations on these animals from diverse analytical and interpretative perspectives were undertaken and developed since the 1980's. These discuss the role of sea lions in human group subsistence during the Middle and Late Holocene (see a synthesis of this research in Muñoz 2011).

By way of contrast, along the continental Patagonian Atlantic littoral, there are very few specialized studies, and quantitative investigations on these *taxa* are scarce (see Table 1). Without doubt, the difficulties in identifying

[1] Good summaries on the history of research along the Patagonian coast exist in Gómez Otero 1995, Orquera & Gómez Otero 2007.

precisely the two species of otariids – the South American fur seal [*Arctocephalus australis* (*Aa*)] and the South American sea lions [*Otaria flavescens* (*Of*)] – commonly found in the archaeofaunal assemblages of southern South America compounded the development of these zooarchaeological studies. Both species of otariids share almost identical post-cranial skeletal morphology and are highly dimorphic; these two factors make it much harder to distinguish between these two *taxa*. The greatest identification problems occur amongst non-fused (immature) individuals, making it difficult to separate them between the female remains of the larger species (*Of*), and males from the smaller one (*Aa*). Legoupil (1989-90) and Schiavini (1990) had already identified these problems by the end of the 1980's in their zooarchaeological studies of pinnipeds amongst marine hunter-gatherers in Tierra del Fuego-Patagonia (Legoupil 1989-1990:107; Muñoz 2011:307-8). Furthermore, most of the zooarchaeological assemblages from the Patagonian littoral came from surface-eroded sites (see below), perhaps contributing to the fact that in many of these studies mention of sea lions was brief, in many cases only acknowledging their presence together with some minimal quantified data such as NISP, MNE, MNI, in some instances even this quantified data was missing (Gómez Otero et al. 2009: 1026).

Advances and Limitations in the Use of Autoecological Knowledge in Zooarchaeological Interpretation

By surreptitiously using pinniped distribution observations of the last century some authors have postulated the presence of South American sea lions – *Of* – in the archaeological record (e.g. Arrigoni & Paleo 1991; Miotti 1998;, L´Heureux & Franco 2002; Castro et al. 2004). Indeed, nowadays *Of* is numerically the most common species along the Argentine Patagonian coast, found in more than 85 rookeries or haul outs. In contrast, the South American fur seal – *Aa* – is restricted to very few rookeries or haul outs, no more than 10 in all Patagonia located on islets and islands to the south of Chubut and in Santa Cruz (Crespo et al. 2007) where they select primordially escarped rocky substrata open to the battering sea (Sielfeld et al. 1978). Thus, archaeological studies along the coasts of Chubut and northern Santa Cruz have noted the presence of both otariids species (*Aa* and *Of*) in the zooarchaeological record. However, these studies do not specify how they distinguished between the two species (*e.g.* Bahía Solano 16, Caviglia et al. 1982:6 MS); in other cases when the data was quantified they were again subsumed under the general heading of pinnipeds (see Table 1) (Moreno 2008; Gómez Otero 2007), and identified sex and age mentioned the use of reference collection without more data (Gómez Otero et al. 2013:180).

Even though we now know that these animals exhibit a certain degree of philopatry (Grandi et al. 2008), we should likewise acknowledge that the present spatial distribution of pinniped colonies was probably influenced by differential rates of recovery by these species in response to the historically recent mass culling of these animals. In this manner, indiscriminate otariid – *Aa* and *Of* – hunting from the 18th through to the mid-twentieth century brought the population of these animals down to critical levels. In addition, increased fishing activity in the last 30 years (Crespo et al. 2007) and other human actions are sufficient in themselves to indicate that the Patagonian marine ecosystem has suffered dramatic changes during this historical period (Balech & Ehrlich 2008). These changes are such that it is impossible to assume similarities or continuities between present and prehistoric otariid spatial distributions. What the extant historical data does prove is that the Patagonian coast has areas suitable for the establishment of sea lions colonies (Borella 2006) and that the existing location of these settlements are a useful *proxy* with which to explore particular aspects of the past (Borrero 2004).

In this manner Moreno and colleagues (2000) analyzed the spatial distribution along the littoral of a lithic artifact type known as the 'skull-crusher'. This was a kind of stone maul used for crushing the skull of an otariid. Examples of this tool-type were recovered mainly from archaeological sites between Cabo Blanco and Puerto Santa Cruz (Northern Santa Cruz). Skull-crushers from museum collections with known provenance were also studied. The authors highlighted that almost all these artifacts came from coastal areas associated with modern sea lions rookeries. Therefore, they concluded that it was highly probable that these tools were used in the hunt of sea lions. Recently these artifacts have been found in other areas of the Patagonian coast in association with high densities of pinniped bones such as at Punta Entrada, south of Santa Cruz (Cruz et al. 2011: 37) and at Punta Pórfido in Nordpatagonia, Río Negro (Borella et al. 2013a)-(see Figure 1). Given the difficulty in finding wood along the coast from which to fashion clubs or garrotes, a common practice in the culling of sea lions documented for other parts of the world (*e.g.* Lyman 1991), these stone skull-crushers would have been an effective alternative technology.

Table 1. Archaeological sites on the Patagonian Atlantic coast with Pinniped remains mentioned in the zooarchaeological record, ordered from North to South.

Archaeological Site	Coastal Area or Location	Species Mentioned	Dates (years BP)	SP/MNI	Sources
Ea. La Serranita (A-)	Bahía San Blas (Buenos Aires)	SNS	5300±50	ND/ND	Sanguinetti 1999:13
Ea. La Eloisa (1)	Bahía San Blas (Buenos Aires)	SNS	1300±40	ND/ND	Sanguinetti 1999:19
Ea. Las Olas (5)	Bahía San Blas (Buenos Aires)	SNS	570±40	ND/ND	Sanguinetti 1999:20
Ea. La Serranita (sector 2)	Bahía San Blas (Buenos Aires)	Otaridae	3690±50	1/1	Eugenio & Aldazabal 2004:696
El Haras sitio 1	Bahía San Blas (Buenos Aires)	Otaridae	3070±70	80/4	Eugenio & Aldazabal 2004:696
Conchero El Lobito	Bahía San Blas (Buenos Aires)	Otaridae	3210±60	18/2	Eugenio & Aldazabal 2004:696
Ea.Olas (Conchero 1)	Bahía San Blas (Buenos Aires)	Otaridae	1960±50	10/1	Eugenio & Aldazabal 2004:696
Ea.Olas (Conchero 5) Sur Ax.	Bahía San Blas (Buenos Aires)	Otaridae	570±40	2/1	Eugenio & Aldazabal 2004:696
Ea.Olas (Conchero 11)	Bahía San Blas (Buenos Aires)	*Otaria flavescens*	2810±50	327/4	Eugenio and Aldazabal 2004:696
Conchero el Piche 1	Bahía San Blas (Buenos Aires)	Otaridae	1500±40	99/2	Eugenio & Aldazabal 2004:696
Bajada de los Pescadores 2	Costa Norte Golfo San Matías (Río Negro)	*A. australis* *O. flavescens*	2197±38 2601±51	133^{-}/ ND (*Aa*)ND /3 (*Of*) ND /1	Borella &Cruz 2012:47
Ea. La Pastosa	Península de Valdés (Chubut)	SNS	ND	ND/ND	Gómez Otero et al. 1999:401
Rincón de Elizalde (2 C 1)	Costa oeste Golfo San Matías (Chubut)	SNS		3 / 2	Gómez Otero 2007:242
Flechero del 39	Golfo San José (Chubut)	SNS	2640±40	ND/ND	Gómez Otero 2007:152
Playa Galván	Golfo San José (Chubut)	SNS	ND	ND/ND	Gómez Otero 2007:153
San Román 3	Golfo San José (Chubut)	SNS	ND	ND/ND	Gómez Otero 2007:155
La Armonia M2	Golfo San Matías, en Península de Valdés (Chubut)	*Otaria* sp	460±40	27/1	Gómez Otero 2007:257
El Progreso 1- MA	Golfo San Matías, en Península de Valdés (Chubut)	*Otaria* sp	1940±60	4/2	Gómez Otero et al. 1999:401, Gómez Otero 2007:259

Table 1. (Continued)

Archaeological Site	Coastal Area or Location	Species Mentioned	Dates (years BP)	SP/MNI	Sources
El Progreso 2- MA	Golfo San Matías, en Península de Valdés (Chubut)	*Otaria* sp	ND	2/1	Gómez Otero et al. 1999:402, Gómez Otero 2007:260
Los Abanicos1 (F 1)	Golfo San Matías, en Península de Valdés (Chubut)	*Otaria* sp. *Arctocephalus* sp.?	380±60	229/3 ND/ND	Gómez Otero & Suarez 2005 (in press); Gómez Otero 2007:260
Punta Pardelas 1 (C1)	Golfo Nuevo (Chubut)	SNS	ND	ND/2	Gómez Otero 2007:270
La Azucena 2 (M1)	Península de Valdés (Chubut)	SNS	ND	2/2	Gómez Otero et al. 1999:400, Gómez Otero 2007:273
Ecocentro 1(F1)	Puerto Madryn, Golfo Nuevo Sur, (Chubut)	*Arctocephalus* sp.?	ND	103/3	Gómez Otero 2007:275
Ecocentro 1(F2)	Puerto Madryn, Golfo Nuevo Sur, (Chubut)	SNS	ND	41/2	Gómez Otero 2007:276
Ecocentro 1(F3)	Puerto Madryn, Golfo Nuevo, Sur (Chubut)	SNS	ND	25/1	Gómez Otero 2007:276
Bahía Cracker 4 (M1)	Golfo Nuevo (Chubut)	*Otaria* sp. *Arctocephalus* sp.	5390+130	10/1 6/1	Gomez Otero et al. 2013:180
Barranca Norte 1 (F1)	Costa norte Estuario Río Chubut (Chubut)	*Otaria* sp. *Arctocephalus* sp.?	1040±70	434/12	Gómez Otero 2007:281
Barranca Norte 1 (C1)	Costa norte Estuario Río Chubut (Chubut)	SNS	3290±80	8/1	Gómez Otero 2007: 282
Barranca Norte 2 (N2)	Costa norte Estuario Río Chubut (Chubut)	SNS	ND	19/1	Gómez Otero 2007: 283
Barranca Norte 6	Costa norte Estuario Río Chubut (Chubut)	SNS	ND	14/1	Gómez Otero 2007: 284
Los Cangrejales Sur Sector 4 L3	Sur del Río Chubut (Chubut)	SNS	1980±60	6/1	Gómez Otero et al. 2009: 1030

Archaeological Site	Coastal Area or Location	Species Mentioned	Dates (years BP)	SP/MNI	Sources
Bahía Solano (BS16)	Golfo San Jorge (Chubut)	*Otaria* sp. and *Arctocephalus* sp.	2954±195	ND/2 ND/1	Caviglia et al. 1982 MS: 5,6
Restinga Alí	Golfo San Jorge (Chubut)	*O. flavescens*	770±60	191/8	Arrigoni & Paleo 1991:207 Arrigoni et al. 2008:98 Moreno 2008: 50
Playa Alsina Sitio III	Golfo San Jorge (Chubut)	*Otaria* sp.	770 *ca*	ND/ND	Arrigoni et al. 2008:100
Cabo Blanco 1	Golfo San Jorge Sur, Norte de Santa Cruz	*Otaria* sp. and *Arctocephalus sp.*	1420±50 1700±30	193/7	Moreno 2008:40, Moreno et al.1998:172
Cabo Blanco 2 (Upper Component) (Lower Component)	Golfo San Jorge Sur, Norte de Norte de Santa Cruz	*Otaria* sp.	960±60 3310±50	71/5 48/4 71/5	Castro et al.2000:558 Moreno 2008:44-45
Médanos del Salitral	Norte de Santa Cruz	SNS	ND	3/1	Moreno 2008:49
Punta Medanosa 1	Norte de Santa Cruz	SNS	2390±90	89/4	Moreno 2008:53
Punta Medanosa 2	Norte de Santa Cruz	SNS	ND	149/8	Moreno 2008:53
Punta Medanosa 3	Norte de Santa Cruz	SNS	2240±80	338/13	Moreno 2008:53
Punta Medanosa 4	Norte de Santa Cruz	SNS	920±40	98/7	Moreno 2008:54
Isla Lobos	Norte de Santa Cruz	SNS	ND	31/3	Moreno 2008:54
Punto 37	Punta Entrada, desembocadura del Río Santa Cruz, Santa Cruz	*A. australis* *O. flavescens*	1540±70	1977¨/ND	Muñoz et al. 2013:462
Punto 96	Punta Entrada, desembocadura del Río Santa Cruz, Santa Cruz	*A. australis* *O. flavescens*	2050±110 1330±100 1759±80	(*Aa*)5/ND (*Of*)18/ND 620¨/11¨ 1112¨/ND	Cruz et al. 2011:35-36, Muñoz et al. 2013: 462
Cabeza de León 1	Monte León, Santa Cruz	SNS	970±100	94¨/3¨	Cruz et al. 2011:35-36
Conchero 4 (CCH4)	Monte León, Santa Cruz	*A. australis* *O. flavescens*	930±60 1430±70	108¨/4¨ (*Aa*) 6/ND (*Of*) 1/ND	Caracotche et al. 2005:148, 157, Cruz et al.2011: 35-36
Localidad 1 CA (Casco)	Punta Bustamante, Cabo Buen Tiempo (Santa Cruz)	*O. flavescens*	ND	6/1	Miotti 1998:184,186

Table 1. (Continued)

Archaeological Site	Coastal Area or Location	Species Mentioned	Dates (years BP)	SP/MNI	Sources
Localidad 2 CE	Punta Bustamante, Cabo Buen Tiempo (Santa Cruz)	*O. flavescens*	ND	5/1	Miotti 1998:192,193
Sitio CE2: sector LPM	Punta Bustamante, Cabo Buen Tiempo (Santa Cruz)	*O. flavescens*	1060±50	15/1	Miotti 1998:198,200 Mansur 2008:176
Sitio 3: Caño	Punta Bustamante, Cabo Buen Tiempo (Santa Cruz)	*O. flavescens*	ND	155/ND	Miotti 1998:203
Sitio RUD0 1BK	Punta Bustamante (Santa Cruz)	*O. flavescens*	3050±60 3690±80	ND/2	Mansur et al. 2004:763, 768
Sitio HST0 1AM	Punta Bustamante (Santa Cruz)	*O. flavescens*	750±70 890±90	22 #/ND	Mansur 2007:702,703
Cabo Vírgenes (1)	Cabo Vírgenes (Santa Cruz)	*O. flavescens*	1380±180	43/2	Barberena et al. 2004:424.426 Borella 2010:125
Cabo Vírgenes (2)	Cabo Vírgenes (Santa Cruz)	*O. flavescens*	1050±70	20/2	Barberena et al. 2004:424,426 Borella 2010:125
Cabo Vírgenes (6)	Cabo Vírgenes (Santa Cruz)	*O. flavescens*	1190±60 1170±50	64/4	L´Heureux & Franco 2002 Barberena et al. 2004:424,426, Borella 2010:125
Cabo Vírgenes (20)	Cabo Vírgenes (Santa Cruz)	*O. flavescens* *A. australis*	1256±50 733±47	(*Aa*) 7/1 (*Of*) 63/4	Belardi et al. 2011: 283 L´Heureux et al. 2012: 240

Legend:

ND: no data

SNS: species not specified

*: Sea elephant

#: Marine mammal¨: pinniped

Figure 1. Skull-crusher from Punta Pórfido, San Matías Gulf, Río Negro Province (Nordpatagonia) Argentina.

EVALUATING ASSEMBLAGE INTEGRITY USING TAPHONOMIC STUDIES

Over the last few decades the process of desertification has affected a large area of coastal Patagonia, conducting to the destruction of the vegetation cover that protected many of the archaeological sites. Increasing aeolian erosion has helped to expose the archaeological record on the surface (Caracotche et al. 2005; Cruz & Caracotche 2008). This has obvious implications concerning what is preserved, and the general state of preservation of the zooarchaeological record. This exposure can also lead to contamination and erroneous associations between archaeological and non-archaeological remains from different origins. The exposure and gradual degradation of archaeofaunal remains underlines the need to carry out taphonomic controls; this should include assessing the formation processes of the assemblage, a task that should commence at the beginning of the archaeological investigation (Borrero 2001). Considering that a large number of the coastal sites with marine vertebrate remains are near-to or on actual ancient and modern coastal shorelines, it is imperative to understand the process by which bone accumulations deposit naturally along these littoral areas (Borella 2004). Taphonomic analysis permits the interpretation of depositional variation and the preservation properties of bones at different places, and provides a means by which to assess the possible presence of naturally deposited bones in archaeological sites.

The joint development of taphonomic and archaeological studies along the continental Patagonian coast commenced towards the end of the 1990's under the direction of Borrero under the remit of the Regional Taphonomy perspective (*sensu* Borrero 2001) at Cabo Vírgenes (Santa Cruz). Nevertheless, his studies there revealed very few remains of pinniped accumulations that could be used to identify the characteristics underlying naturally deposited coastal assemblages.

There were other taphonomic studies undertaken on Patagonian littoral environments but these were focused on other *taxa* such as marine birds (Moreno 2008, and for a synthesis of bird taphonomic studies see Cruz 2011) or on very specific case studies. One such study was the taphonomic analysis of various pinniped skeletons from the 'Lobos site' (Chubut). The bone assemblage was from a single episode of natural mass death of 22 *Of* individuals dated to 1,200 BP (Serrán et al. 2008).

Nonetheless, to recap there have been no systematic taphonomic studies on pinniped remains documenting the composition of modern bone accumulations; nor have there been zooarchaeological studies directed at revealing the exploitation strategies of sea lions themselves, even though they were a crucial subsistence resource for hunter-gatherers, and their remains are found in the majority of coastal sites of continental Patagonia.

It was this scenario that led to the development of methods geared to resolving the problems inherent in identifying these *taxa*. The objective being to generate better and more precise data that would in turn permit a thorough reassessment of the role played by pinnipeds amongst the hunter-gatherers of Patagonia.

New Perspectives: In the Pursuit of Criteria to Evaluate the Pinniped Zooarchaeological Record

In the last few years, new areas of archaeological research opened up along the coast of Patagonia and with these the first specialized studies targeting the large and abundant pinniped zooarchaeological remains. It is in these studies that we found a first mention of *Aa* remains along sectors of the continental coast, where at present no rookeries or haul outs of these animals exist, such as Nordpatagonia (Borella 2006) or southern Santa Cruz Province, at the Monte León National Park (Caracotche et al. 2005) and at Cabo Vírgenes (Belardi, Carvallo Marina & L'Heureux 2011; L'Heureux, Belardi & Carvallo Marina 2012).

Nevertheless, the identification of *Aa* maxillaries and teeth alerted us to the unequivocal presence of this species along the continental Patagonian coast during the Late Holocene. In Nordpatagonia, on the coast of the San Matías Gulf, the majority of the archaeological sites with abundant *Aa* remains were found in littoral areas without off-coast islands or islets. These are exactly the types of locales favored by this species in the present for their rest and breeding activities. The archaeological evidence would seem to suggest that in the past there were a greater number of *Arctocephalus australis* along the Patagonian continental coast than is the case today. Also these ancient *Aa* rookeries were situated along the continental coast in habitats similar to those occupied by this taxon at present.

Yet the Nordpatagonia coastal data raised various questions, such as, what was the proportional representation of the two separate otariids species in the archaeological record? Were they hunting for a particular prey, or selecting for certain age profiles or size? Was there a change in the exploitation of these animals during the period in which this coastal area was occupied? In order to answer these questions it would first be important to improve the quality of the available data. This need to improve the data, would in turn necessarily lead to the development of adequate methods with which to overcome the old limitations mentioned above.

The Development of Osteological Methods for the Zooarchaeological Analysis of Otariid Remains

In other parts of the world, inter-specific identification of pinnipeds was undertaken using a combination of different analytical methods, such as: 1- identify morphologically particular bones that allowed for species identification; 2- know the chronology of the epiphyseal fusion sequence; and, 3- carry out osteometric studies on modern reference collections with a view to using these measurements comparatively when analysing the archaeological remains, which in turn allows to distinguish according to species and age profiles (e.g. Hodgetts 1999; Storå 2001, 2002).

Regarding the first point above, to date there is only one very preliminary study that deals with this issue. This study was undertaken on seven samples from the Uruguayan coast and it sought to develop identification criteria for *Aa* and *Of* post-cranial skeletons. Given that the species studied in this research were the same as those in Patagonia its comparative importance was readily apparent. By using a small sample size of reference (four *Aa*, but only two complete males, one male *Arctocephalus tropicalis*, and two complete male *Of*) Pérez García (2003) describes different characteristics on a number of bones (scapula, humerus, radius, ulna, innominate, femur, tibia, fibula, calcaneum, atlas and axis). Nevertheless, as stated by the author, the sample size would have to be increased and include more females to validate these results.

In a study looking for traits to differentiate amongst the several species of North Atlantic phocids (seals) found in the archaeological record, Hodgetts (1999) suggested that it would be better to study seven or more individuals from each of these species, given the existing variation within each species. The study concluded that not all bones were useful species indicators, so that a number of traits, rather than one in particular, had to be employed during identification. The points raised by Hodgetts (1999) are currently being assessed by research in progress (Borella et al. 2013b).

OSTEOMETRIC STUDIES

The presence of pinniped pups in the zooarchaeological record has been used as an indicator of seasonality for various coastal sites along southern Patagonia (Miotti 1998; L´Heureux & Franco 2002; Lefevre et al. 2003 amongst others). This assumption is based on the fact that both otariid species give birth on land at a particular moment in the year, that being during the summer. Thus, the presence of very small pinniped individuals (non-fused or immature bones) in the archaeological record would represented summer captures. Yet, there was a serious missive in these studies, being that in none of these cases the pup bones were identified as newborn.

However, metric data on yearling bones can be used to give an estimate of the season during which death occurred. Borella and L'Heureux (2010) have conducted the first osteometric studies along the Patagonian coast undertaken on modern reference collections of young *Otaria flavescens* of known age and sex. Their aim was to determine from a study of yearlings possible indicators of seasonality. In this vein, the metric data obtained demonstrated that the maximum length of the bones was a good measure for assessing size range amongst pups. Bone length also permitted the identification of newborn, and thus provided a good indicator for determining summer occupation in archaeological sites (Borella & L´Heureux 2010; Borella et al. 2010). Nevertheless, and according to Monks (1981), in order to infer seasonality of capture, it may be better to use several indicators to obtain testable hypotheses (Borella et al. 2014).

In coastal sites of the San Matías Gulf, Río Negro, there is a preponderance of non-fused – young and sub-adult – pinniped bones (Borella 2006). Assuming that modern populations of sea lions are more similar in body size than those encountered in the archaeological record, Borella and L'Heureux (2013) are employing osteometric analysis on the zooarchaeological pinniped remains in order to identify species and type of prey in the Nordpatagonia record. On this basis, we are identifying what proportion of both species are represented over time and if there was a change – or not – in the exploitation strategies of sea lions during the Late Holocene. Stable isotope analysis on human samples from this area of the Nordpatagonia coast showed two types of palaeo-diets that followed each other chronologically: a predominately marine diet between 6,000 and 2,200 BP, probably already in place before 6,000 BP, and a mixed diet between 1,500 and 420 BP (Favier Dubois et al. 2009). Without doubt ecological isotope analysis is a useful tool in determining which, when and where marine resources were being exploited. It is also possible to study whether there were changes in the exploitation of marine resources through time via the detailed analysis of the composition of otariid remains in the zooarchaeological record. Borella and L´Heureux's osteometric study on six sites from different periods showed a tendency towards a predominance of *Aa* over *Of* in the analyzed samples. From this we infer that *Aa* was much more abundant in the past that it is at present (Borella & L´Heureux 2013a, 2013b).

BONE FUSION IN SOUTH AMERICAN OTARIIDS

In order to develop osteometric studies on South American otariids it was first necessary to compile a reference collection with known age and sex profiles,[1] this in turn allowed us to identify the epiphyseal fusion sequence in both species. From these studies, Borella et al. (2013c) has been able to propose a preliminary time sequence for bone fusion of post-cranial elements in both species (*Aa* y *Of*).

Even if this preliminary data is somewhat incomplete -given that the study was conducted on an assemblage with a natural mortality profile and therefore not all age classes were represented-, it still permitted a first approximation into the time taken for bone fusion of post-cranial elements in these two otariid species. Bone fusion amongst these *taxa* coalesced into two groups; an early fusion group – distal humerus, proximal epiphyses of the radius, proximal epiphyses of the tibia/fibula, and others – and a late fusion group – ulna, radius distal, tibia/fibula distal and calcaneus tuberosity (Borella et al. 2013d).

Differential Preservation of Pinniped Skeletal Parts

A study of bone mineral density values of these two species (*Aa* and *Of*) was undertaken to assess the reliability of the zooarchaeological samples. This data provided an indication of the preferential preservation of certain skeletal elements from these animals and suggested possible biases within the faunal record. We therefore studied the mineral bone density from the post-cranial skeletons of adult males, selecting one sample from each of these two otariid species. The bone density method employed was the same as that used on the guanaco and ñandú, two other species crucial to human subsistence in this region of South America (Borella et al. 2007).

Even so, a bone mineral density study only addressed part of the criteria determining bone assemblage formation given that the physical context of the assemblage itself is important. It is therefore necessary to consider bone mineral density values in combination with other types of data such as taphonomy. In taking this last point into consideration, with the added aim of understanding the characteristics underlying the present-day bone accumulations found near to *Of* breeding rookeries in Nordpatagonia -Islote

[1] Age and sex determination of all samples were undertaken at the Laboratorio de Mamíferos Marinos CENPAT-CONICET under Dr. E.A. Crespo. The collection of *Otaria flavescens* young of less than a year of age were recovered from the Golfo San Matías coast (Río Negro, Argentina) and deposited at the INCUAPA-CONICET (UNICEN). The samples belonging to individuals older than a year from both species belong to the LAMAMA-CENPAT-CONICET collection.

Lobos (Río Negro Province)-, Borella and Borrero (2010) developed a series of taphonomic observations regarding the disarticulation pattern of pinniped carcasses; the rates of bones occurrence per m^2; the degree of bone completeness and the visibility of pup bones in natural records along the shoreline. To the south of the Santa Cruz River, Santa Cruz Province, along another coastal stretch of continental Patagonia, recent research uncovered considerable remains of both pinniped species. These include the archaeological deposits from Punta Entrada and Monte León with a chronological spread covering the last 2,000 years (Cruz et al. 2010; Cruz et al. 2011; Muñoz et al. 2013). These researchers have worked in tandem the taphonomy and zooarchaeology of this area since the start of their project – since 2005 – with the aim of understanding the formation process of the bone accumulations they have registered. Their taphonomic observations have looked at different *taxa* (terrestrial and marine), and even though their results on sea lions remains have yet to be published we are aware of their research agenda. This research agenda involves understanding the process of decomposition and disarticulation in different environments, an assessment of weathering and the formation of natural pinniped faunal assemblages, as well as the comparative study of the impact of diverse taphonomic agents on different *taxa* focusing on longitudinal observations of skeletons (*sensu* Haynes 1995:9). These results were presented at recent scientific meetings (Muñoz & Cruz 2010, 2011, 2013).

Specific Archaeofaunal Analysis on These Taxa

There are still very few published articles on the results of archaeofaunal analysis of pinniped remains from coastal continental Patagonia. Indeed the first group to provide information on the analysis of otariid remains are those working at the sites of Punta Entrada (Cruz et al. 2011; Muñoz et al. 2013), and Cabeza de León 1 (Muñoz 2015). In particular at Punta Entrada there was a preponderance of pinniped remains in respect to other marine resources, in this case it comprised more than 75% of the taxonomic assemblage (Cruz et al. 2010; Cruz et al. 2011; Muñoz et al. 2013). Given that the material was recovered from a surface context an assessment was undertaken on the reliability of the assemblage by examining the degree of bone fragmentation, as well as the possible link between weathering and the differential preservation of otariid bones using bone mineral density. They concluded that there was no direct correlation in these samples between these two last variables. On the basis of taphonomic observations in this area, these researchers concluded that weathering of otariid bones was slower than that of other *taxa*.

Another important aspect of this analysis was the researchers presentation and discussion of the anatomical representation of skeletal elements[2] and the differences in respect to the number of individuals (MNE/MNI) present in these archaeological assemblages (Cruz et al. 2011; Muñoz et al. 2013). Likewise they present a study of the locations of cuts and other anthropogenic modifications that occurred on pinniped cortical bone surface, with a view to identifying specific dismemberment and butchering procedures. Both these concerns are central to our understanding of the processing of prey and the transportation of skeletal elements, which in turn helps to delineate the underlying exploitation methods of these animals, an aspect that had not been addressed by previous studies in continental Patagonia (Borrero 2004).

Finally, we should mention the discovery of two mono-dentate bone harpoons from a shell midden at Punta Entrada (Santa Cruz). These harpoons were morphologically similar to those studied by Moreno (2008) from the Punta Medanosa area, from northern Santa Cruz (Cruz et al. 2011). These tools, generally made from marine mammal bones, are rare in continental Patagonia and more often than not are found in private collections or museums. Given that these tools have been found in conjunction with pinniped remains, they have been interpreted as artifacts for use in hunting aquatic or marine prey (Moreno 2008), and together with the aforementioned 'skull-crusher' mauls, were probably used for hunting otariids. Cruz et al. (2011) stated that even given the differences in size between the two species – *Of* are bigger than *Aa*– hunting of both otariids was conducted with similar, or even the same tools.

CONCLUSION

The particular characteristics of the pinniped zooarchaeological record and its archaeological recovery along the Patagonian coast have led different researchers into adopting diverse insights into the nature of their finds and their depositional context. Yet a deeper understanding of the types of

[2] The archaeofaunal analysis of the Cabo Vírgenes 20 site briefly mentions the skeletal elements of the pinnipeds present (Belardi et al. 2011:284), and identified MNI for *Aa* and *Of,* but do not specify how they distinguished between the two species (see Belardi et al. 2011: 284; L´Heureux et al. 2012:240).

exploitation of sea lions along the Patagonian littoral will depend on the progress of our taxonomical assessments. As stated by Lyman (2011:29) '*Correct taxonomic identification is critical to any zooarchaeological endeavour*'. Therefore, a lack of detailed identification and analysis of pinnipeds from the archaeological record makes our interpretations weak, especially when assessing the role that these animals played in the subsistence strategies of past human populations. Nowadays we should not content ourselves merely with noting their presence and the difficulties in correctly identifying them, especially if we strive to more fully understand the use of space and coastal resources by past hunter-gatherers of these lower latitudes. Fortunately in the last few years, the rise in archaeological research on the Patagonian coast has produced a considerable amount of zooarchaeological data on these *taxa* paralleled by the development of methods targeted at studying these remains. This invites us to explore questions such as, what was the origin of their (pinnipeds) presence? How were they exploited (strategies used)? Since when were they exploited and were there changes in these exploitation strategies through time?

As has been seen, our reflections of a large number of the zooarchaeological studies considered here seek to resolve the limitations of the samples studied. In this, taphonomic studies, methods and reference frameworks provide a crucial pivot with which to address specific problems (representation of elements and skeletal parts, differential preservation, bone fusion chronology, vertical movement of bones, taphonomic characteristic of naturally deposited bones, etc.) whilst at the same time permitting the fine-tuning of our research objectives. Even though pinniped research has taken great strides forward, on certain aspects there is still a long way to go (see discussion in Muñoz 2011). Nevertheless, the future looks promising, as stated by Borrero (2011:179):

> "We work in a discipline for which most of our samples are affected in one way or another, and acquiring knowledge about the limitations of our data is the best way to pursue our research goals."

Zooarchaeological studies on pinniped remains has greatly developed over the last few decades in different parts of the world, and many of the methodological problems mentioned in this chapter have been acknowledged and resolved by these other studies. European studies into pinniped exploitation in the Baltic sea (Storå 2001), or in Arctic Norway (Hodgetts 1999); as well as along the West Coast of North America (Lyman 1995, 2011), or New Zealand (Smith 2005), can provide possible comparative scenarios and alternative insights that can enrich our understanding of the variability of exploitation strategies vis-à-vis these animals.

The challenge for the future will be the use of these and other sources, in an effort to acquire ever more precise data; this will have to be allied to the appropriate quantification methods so that the end result are zooarchaeological analysis in tune with the research questions postulated. This will allow us to address the twin goals of modern zooarchaeology – understand the biology and ecology of animals and that of human behavior through space and time (Reitz & Wing 1999).

REFERENCES

Arrigoni, G. I. & Paleo, M. C. (1991). Investigaciones Arqueológicas en la región Central del Golfo San Jorge, *Shincal 3(3)*, 206-210.

Arrigoni, G. I., Andrieu, M. & Bañados, C. (2008). Arqueología de cazadores-recolectores prehistóricos en la costa central del Golfo San Jorge. In Cruz, I. & M. S. Caracotche (Eds.). *Arqueología de la costa patagónica: perspectivas para la conservación* (pp. 91-107). Río Gallegos, Argentina: Universidad Nacional de la Patagonia Austral - Subsecretaría de Cultura de la Provincia de Santa Cruz.

Balech, E. & Ehrlich, M. D. (2008). Esquema Biogeográfico del Mar Argentino. *Revista de Investigación y Desarrollo Pesquero 19*, 45-75.

Belardi, J. B. Carballo Marina, F. & L'Heureux G. L. (2011). Nuevos Resultados Arqueológicos en Cabo Vírgenes (Santa Cruz): El sitio Cabo Vírgenes 20. *Magallania 39*, 279-292.

Borella, F. 2004. *Tafonomía Regional y Estudios Arqueofaunísticos de cetáceos en Tierra del Fuego y Patagonia Meridional, British Archaeological Report (BAR 1257).* Oxford, UK: Archaeopress.

Borella, F. (2006). ¿Dónde están los lobos en la costa atlántica de Norpatagonia? Explorando vías para resolver el registro arqueofaunístico. *Werken 9*, 97-114.

Borella, F. & Borrero, L. A. (2010). Observaciones tafonómicas acerca de la desarticulación de carcasas de pinnípedos en ambientes litorales, el caso de islote Lobos (Golfo San Matías, Río Negro). In Gutiérrez, M., M. De Nigris, P. Fernández, M. Giardina, A. Gil, A. Izeta, G. Neme & H. Yacobaccio (Eds.). *Zooarqueología a principios del siglo XXI: aportes teóricos, metodológicos y casos de estudio* (pp. 371-379). Buenos Aires, Argentina: Ediciones del Espinillo.

Borella, F. & L'Heureux G. L. (2010). Primeros resultados osteométricos para una discusión de la estacionalidad de captura de otáridos en sitios arqueológicos de Patagonia: el caso de Cabo Vírgenes (Santa Cruz). In Bárcena, R. & H. Chiavazza (Eds.). *Arqueología Argentina en el Bicentenario de la Revolución de Mayo* (pp. 1865-1870). Mendoza, Argentina: FFyL UNCuyo & ICHSA CONICET.

Borella, F. & L'Heureux G. L. (2013). Determinando Presas: primeros resultados osteométricos para la identificación de especies de otáridos en Concheros de Nordpatagonia (Río Negro, Argentina). *Revista Chilena de Antropología*. In press.

Borella, F. & L'Heureux G. L. (2013). Explorando variaciones corológicas y morfológicas de los principales mamíferos explotados en una localidad costera nordpatagónica del Holoceno tardío. In *Libro de resúmenes del III Congreso Nacional de Zooarqueología* (p. 35). Buenos Aires, Argentina: Facultad de Filosofía y Letras, Universidad de Buenos Aires.

Borella, F. & Cruz, I. (2012). Taphonomic evaluation of penguin (Spheniscidae) remains at a shell-midden on the northern coast of Patagonia (San Matías Gulf, Río Negro, Argentina). *Quaternary International 278*, 45-50.

Borella, F., Grandi M. F. & L'Heureux G. L. (2010). Evaluation of summer exploitation of *Otaria flavescens* in coastal archaeological sites of Patagonia. In *Abstracts ICAZ 11th. Conference* (p. 99). Paris, France.

Borella, F., L'Heureux G. L. & Grandi, M. F. (2014). Osteometric analysis of South American Sea Lions (*Otaria flavescens*) pups from Patagonia. An Assessment of their use as indicators for seasonality in archaeological sites. *International Journal of Osteoarchaeology,* DOI 10.1002/oa 2392.

Borella, F., Gutíerrez, M. A., Fodere, H. R. & Merlo J. F. (2007). Estudio de densidad mineral ósea para dos especies de otáridos frecuentes en el registro arqueofaunístico patagónico. In Morello, F., M. Martinic, A. Prieto & G. Bahamonde (Eds.). *Arqueología de Fuego-Patagonia. Levantando piedras, desenterrando huesos... y develando arcanos* (pp. 421-426). Punta Arenas, Chile: Ediciones CEQUA.

Borella, F., Grandi, M. F., Vales, D. G., Goodall R. N. P. & Crespo, E. A. (2013). Esquema preliminar de fusión epifisaria en huesos de lobos marinos (*Arctocephalus australis* y *Otaria flavescens*), su contribución en los análisis zooarqueológicos. In A. F. Zangrando, R. Barberena; A.Gil, G.Neme, M. Giardina, L. Luna, C. Otaola, S. Paulides, L. Salgán & A.

Tivoli (Eds.). *Tendencias teórico-metodológicas y casos de estudio en la arqueología de Patagonia* (pp. 39-51). San Rafael, Argentina: Museo de Historia Natural de San Rafael.

Borella, F., Cardillo, M., Favier Dubois, C. M., Scartascini, F. L., Alberti, J., Marani, H. & Borges Vaz, E. (2013a). Las ocupaciones humanas entre Punta Pórfido y Punta Odriozola, Costa Oeste del Golfo San Matías. Nuevos hallazgos y perspectivas. In Bárcena, R. & S. E. Martín (Eds.). *Arqueología Argentina en el Bicentenario de la Asamblea Constituyente del Año 1813* (p.517). La Rioja, Argentina: Universidad Nacional de la Rioja.

Borella, F., Vales, D. G., Grandi, M. F., García, N. A., Crespo E. A. (2013b). Rasgos morfológicos en huesos del postcráneo para la identificación de otáridos en el registro zooarqueológico. In *Libro de resúmenes del III Congreso Nacional de Zooarqueología* (p. 34). Buenos Aires, Argentina: Facultad de Filosofía y Letras, Universidad de Buenos Aires.

Borrero, L. A. (2001). Regional Taphonomy: Background Noise and the Integrity of the Archaeological Record. In Kuznar, L. A. (Ed.). *Ethnoarchaeology of Andean South America. Contributions to Archaeological Method and Theory, R. Whallon* (Series Ed.), *Ethnoarchaeology series, 4* (pp. 243-254). Ann Arbor; USA: International Monographs in Prehistory.

Borrero, L. A. (2004). Tafonomía Regional: el caso de los pinnípedos. In Civalero, M. T., P. M. Fernández & A. G. Guráieb (Eds.). *Contra viento y Marea. Arqueología de Patagonia* (pp. 445- 454). Buenos Aires, Argentina: Instituto Nacional de Antropología y Pensamiento Latinoamericano.

Borrero, L. A. (2011). The archaeology of transformation. *Quaternary International 245*, 178-181.

Caracotche, M. S; Cruz; I., Espinosa S., Carballo F. & Belardi, J. B. (2005). Rescate arqueológico en el Parque Nacional Monte León (Santa Cruz, Argentina). *Magallania 33 (2)*, 143-163.

Castro, A., Moreno, E. J., Martinelli K. & Pepe, F. (2000). Restos faunísticos, artefactos líticos: más información sobre la costa norte de Santa Cruz. In Belardi, J. B., F. Carvallo Marina & S. Espinosa (Eds.). *Desde el país de los gigantes. Perspectivas arqueológicas en Patagonia, v II* (pp. 551-561). Río Gallegos, Argentina: Universidad Nacional de la Patagonia Austral.

Castro, A., Gómez Otero, J., Arrigoni G. & Moreno, J. E. (2004). Prospección macrorregional comparativa a las loberías de la costa atlántica Continental de Patagonia: algunas claves sobre el uso del espacio y de otros recursos.

In Belardi, J. B., F. Carvallo Marina & S. Espinosa (Eds.). *Desde el país de los gigantes. Perspectivas arqueológicas en Patagonia, vol I* (pp. 197-215). Buenos Aires, Argentina: Instituto Nacional de Antropología y Pensamiento Latinoamericano.

Caviglia, S., Borrero, L. A., Casiraghi, M., García L. C. & Horwitz, V. (1982). Nuevos sitios arqueológicos para la región de Bahía Solano (Chubut). *Paper presented at VII Congreso Argentino de Arqueología Argentina*, San Luis, Argentina. M.S.

Crespo, E. A.; Lewis, M. N. & Campagna, C. (2007). Mamíferos marinos: pinnípedos y cetáceos. In Carreto J. I. & C.Bremec (Eds.). *El Mar Argentino y sus recursos pesqueros, INIDEP special publication, v. 5* (pp.127- 150). Mar del Plata, Argentina.

Cruz, I., (2011). Tafonomía de huesos de aves. Estado de la cuestión y perspectivas desde el sur del Neotrópico. *Antípoda, Revista de Antropología 13*, 147-174.

Cruz, I. & Caracotche, M. S. (Eds.) (2008). *Arqueología de la costa patagónica: perspectivas para la conservación*. Río Gallegos, Argentina: Universidad Nacional de la Patagonia Austral - Subsecretaría de Cultura de la Provincia de Santa Cruz.

Cruz, I. & Muñoz, A. S. (2011). Meteorización de restos óseos de guanaco y lobo marino: cambio y estabilidad en carcasas y conjuntos actuales en Punta Entrada y Monte León (costa atlántica de Patagonia). *Libro de resúmenes II Congreso Nacional de Zooarqueología Argentina* (p. 26.). Olavarria, Argentina: FACSO-UNICEN.

Cruz, I., Muñoz A. S. & Lobbia, P. A. (2010). Zooarqueología al sur del río Santa Cruz (Patagonia Argentina). Los restos de fauna del P 96 (Punta Entrada) y CL 1 (P. N. Monte León). In Bárcena, R. & H. Chiavazza (Eds.). *Arqueología Argentina en el Bicentenario de la Revolución de Mayo* (pp. 315-320). Mendoza, Argentina: FFyL UNCuyo & ICHSA CONICET.

Cruz, I., Muñoz A. S. & Lobbia, P. A. (2011). La explotación de recursos marinos en la costa de Patagonia continental: los restos de vertebrados en depósitos de Punta Entrada y Monte León (Santa Cruz, Argentina). *Revista de estudios marítimos y sociales 4*, 31-41.

Eugenio, E. & Aldazabal, V. B. (2004). Los cazadores recolectores del litoral marítimo del área de Bahía de San Blas, Pcia. de Buenos Aires. In Civalero, M. T., P. M. Fernández & A. G.Guráieb (Eds.). *Contra viento y Marea. Arqueología de Patagonia* (pp. 687-700). Buenos Aires,

Argentina: Instituto Nacional de Antropología y Pensamiento Latinoamericano.

Favier Dubois, C. M. ; Borella, F. & Tykot, R. H. (2009). Explorando tendencias en el uso humano del espacio y los recursos en el litoral rionegrino (Argentina) durante el Holoceno medio y tardío. In Salemme, M., F. Santiago, M. Álvarez, E.Piana, M.Vázquez, & M. E. Mansur (Eds.). *Arqueología de la Patagonia. Una mirada desde el último confín T2* (pp. 985-997). Ushuaia, Argentina: Editorial Utopías.

Gómez Otero, J. (1995). Bases para una arqueología de la Costa Patagónica Central (entre el Golfo San José y Cabo Blanco). *Arqueología 5,* 61-103.

Gómez Otero, J. (2007). *Dieta, uso del espacio y evolución en poblaciones cazadoras-recolectoras de la costa centro-septentrional de Patagonia durante el Holoceno medio y tardío.* Unpublished PhD dissertation. Facultad de Filosofía y Letras, Universidad de Buenos Aires.

Gómez Otero, J. & Suárez, F. (2005). Lobos marinos y guanacos: análisis arqueofaunístico de un fogón hallado en la costa del Golfo San Matías, Península de Valdés (Chubut). In *Actas del XIII Congreso Nacional de Arqueología Argentina.* Cabildo Histórico de la Cuidad de Córdoba, Córdoba. In press.

Gómez Otero, J., Weiler, N. & Moreno, J. E. (2009). La localidad arqueológica Los Cangrejales Sur: evidencias de ocupaciones humanas y de variaciones en la línea de costa durante el Holoceno tardío. In Salemme, M., F. Santiago, M. Álvarez, E. Piana, M.Vázquez, & M. E. Mansur (Eds.). *Arqueología de la Patagonia. Una mirada desde el último confín* T2 (pp. 1023-1036). Ushuaia, Argentina: Editorial Utopías.

Gómez Otero, J., Belardi, J. B., Súnico A. & Taylor, R. (1999). Arqueología de cazadores-recolectores en Península Valdés (costa Central de Patagonia): Primeros resultados. In *Soplando en el viento... Actas de las terceras Jornadas de Arqueología de la Patagonia* (pp. 393-417). Neuquen, Argentina: Universidad del Comahue, Facultad de Humanidades e INAPL.

Gómez Otero, J., Weiler, N., Banegas A. & Moreno, J. E. (2013). Ocupaciones del Holoceno medio en Bahía Craker, Costa Atlántica de Patagonia central. In A. F. Zangrando, R. Barberena, A.Gil, G. Neme, M. Giardina, L. Luna, C. Otaola, S. Paulides, L. Salgán & A. Tivoli (Eds.). *Tendencias teórico-metodológicas y casos de estudio en la arqueología de Patagonia* (pp. 177-186). San Rafael, Argentina: Museo de Historia Natural de San Rafael.

Grandi, M., F., Dans, S. L. & Crespo, E. A. (2008). Social Composition and Spatial Distribution of Colonies in an Expanding Population of South American Sea Lions. *Journal of Mammalogy 89 (5)*, 1218-1228.

Haynes, G. (1995). Pre-Clovis and Clovis Megamammals: A Comparison of Carcass Disturbance Age profiles, and Other Characteristics in Light of Recent Actualistic Studies. In Johnson, E., (Ed.). *Ancient Peoples and Landscapes* (pp. 9-27). Lubbock, USA: Museum of Texas Tech University.

Hodgetts, L. M. (1999). *Animal Bones and Human Society in the late Younger Stone Age of Artic Norway*. Unpublished Ph. D thesis. Departament of Archaeology, University of Durham.

Lefevre, C., Lepetz, S. & Legoupil, D. (2003). ¿Cazadores terrestres, cazadores marítimos? Explotación de los recursos animales en el *Locus* 1. In D. Legoupil (Ed.), Cazadores –recolectores del Ponsonby (Patagonia Austral) y su paleoambiente desde VI al III milenio A.C. *Magallania 31*, 63-116.

Legoupil, D. (1989-1990). La identificación de los mamíferos marinos en los sitios canoeros de Patagonia; problema y constataciones. *Anales del Instituto de la Patagonia 19*, 101-113.

L'Heureux G. L. & Franco, N. V. (2002). Ocupaciones humanas en el área de Cabo Vírgenes (Pcia. de Santa Cruz, Argentina): El sitio Cabo Vírgenes 6. *Anales del Instituto de la Patagonia 30*, 183-201.

L'Heureux, G.L., Belardi, J. B., & Carballo Marina, F. (2012). Exploitation of fauna at Cabo Vírgenes, Patagonia Argentina, during the late Holocene. In *Proceedings of the General Session of the 11th International Council for Archaeozoology Conference*. British Archaeological Report (BAR 2354) (pp. 237-244). Oxford, UK: Archaeopress.

Lyman, R. L. (1991). Subsistence Change and Pinniped Hunting. In Stiner, M. C. (Ed). *Human Predators and Prey Mortality* (pp. 187-199). Boulder, USA, Westview Press.

Lyman, R. L. (1995). On the Evolution of Marine Mammal Hunting on the West Coast of North America. *Journal of Anthropological Archeology 14*, 45-77.

Lyman, R. L. (2011). A History of Paleoecological Research on Sea Otters and Pinnipeds of the Eastern Pacific Rim. In Braje, T. J. & T. C. Rick, (Eds). *Human Impacts on Seals, Sea Lions, and Sea Otters: Integrating Archaeology and Ecology in the Northeast Pacific* (pp.19-40). California, USA: The Regents of the University of California.

Mansur, M. E. (2007). Confección y uso de artefactos discoides en contextos de cazadores-recolectores de Patagonia Meridional: Pesas para Redes en el sitio HST01AM (Prov. Santa Cruz, Argentina). In Morello, F., M. Martinic, A. Prieto & G. Bahamonde (Eds.). *Arqueología de Fuego-Patagonia. Levantando piedras, desenterrando huesos... y develando arcanos* (pp. 709-721). Punta Arenas, Chile: Ediciones CEQUA.

Mansur, M. E. (2008). Arqueología de la zona de Punta Bustamante (Prov. de Santa Cruz, Argentina). In Cruz, I. & M. S. Caracotche (Eds.). *Arqueología de la costa patagónica: perspectivas para la conservación* (pp.173-193). Río Gallegos, Argentina: Universidad Nacional de la Patagonia Austral - Subsecretaría de Cultura de la Provincia de Santa Cruz.

Mansur, M. E., Lasa, A. & Vázquez, M. (2004). Investigaciones Arqueológicas en Punta Bustamante, Prov. de Santa Cruz: El sitio RUDO 1BK. In Civalero, M. T., P. M. Fernández & A. G.Guráieb (Eds.). *Contra viento y Marea. Arqueología de Patagonia* (pp. 755-774). Buenos Aires, Argentina: Instituto Nacional de Antropología y Pensamiento Latinoamericano.

Miotti, L. (1998). Zooarqueología de la meseta central y costa de Santa Cruz. Un enfoque de las estrategias adaptativas aborígenes y los paleoambientes. *Revista del Museo de Historia Natural de San Rafael TX (1/4).* San Rafael, Argentina: Museo Municipal de Historia Natural.

Monks, G. G. (1981). Seasonality Studies. *Advances in Archaeological Method and Theory 4*, 177-240.

Moreno, J. E. (2008). *Arqueología y etnohistoria de la Costa Patagónica Central en el Holoceno Tardío.* Rawson, Argentina: Fondo Editorial Provincial, Secretaría de Cultura de Chubut.

Moreno, J. E. Castro, A., Martinelli, K. & Abello, A. (1998). El material faunístico del sitio Cabo Blanco 1. *Anales del Instituto de la Patagonia 26*, 169-179.

Moreno, J. E., Castro, A. & Pepe, F. (2000). El rompecráneos: un artefacto probablemente destinado para la caza de pinnípedos, en la costa de Patagonia continental. In Belardi, J. B., F. Carvallo Marina & S. Espinosa (Eds.). *Desde el país de los gigantes. Perspectivas arqueológicas en Patagonia v II* (pp. 563-572). Río Gallegos, Argentina: Universidad Nacional de la Patagonia Austral.

Muñoz, A. S., (2011). Human-pinniped relationships in Southern Patagonia. Current issues and future research agenda. In Bicho, N., J. Haws & L.

Davis (Eds.). *Trekking the Shore: Changing Coastlines and the Antiquity of Coastal Settlemen* (pp. 305-332). Amsterdam, Holland: Springer.

Muñoz, A. S. (2015). El registro zooarqueológico del Parque Nacional Monte león (Santa Cruz, Argentina): una perspectiva desde el sitio Cabeza de León 1. *Arqueología 21*, in press.

Muñoz, A. S. & Cruz, I. (2010). Taphonomic observations on modern pinniped carcasses and bone assemblages from Southern Patagonia coast", In *Abstracts 11th Conference of the International Council for Archaeozoology (ICAZ)* (p. 192). Paris, France: Muséum National d´Histoire Naturelle.

Muñoz, A. S. & Cruz, I. (2013). Investigaciones tafonómicas naturalistas en la costa atlántica de Patagonia Meridional: estado actual y perspectivas. *Revista Chilena de Antropología*, In press.

Muñoz, A. S., Cruz I., Lemaire, C. R. & Pretto, A. (2013). Los restos arqueológicos de pinnípedos de la desembocadura del río Santa cruz (Punta Entrada, costa atlántica de Patagoni) en perspectiva regional. In A. F. Zangrando, R. Barberena, A.Gil, G. Neme, M. Giardina, L. Luna, C. Otaola, S. Paulides, L. Salgán & A. Tivoli (Eds.). *Tendencias teórico-metodológicas y casos de estudio en la arqueología de Patagonia* (pp. 459-467). San Rafael, Argentina: Museo de Historia Natural de San Rafael.

Orquera L. A. & Gómez Otero, J. (2007). Los Cazadores –Recolectores de las costas de Pampa, Patagonia y Tierra del Fuego. *Relaciones de la Sociedad Argentina de Antropología XXXII*, 75-99.

Pérez García, M. I. (2003). Osteología comparada del esqueleto postcraneano de dos géneros de Otariidae del Uruguay. *Boletín de la Sociedad Zoológica del Uruguay (2º época) 14*, 1-16.

Reitz, E. J. & Wing, E. S. (1999). *Zooarchaeology*. Cambridge, UK: Cambridge University Press.

Sanguinetti de Bórmida, A. C. (1999). Proyecto Norpatagonia. Arqueología de la Costa Septentrional *Anales de la Academia Nacional de Ciencias de Buenos Aires (Anticipo)*, 3-35.

Schiavini, A. C. M. (1990). *Estudio de la relación entre el hombre y los pinnípedos en el proceso adaptativo humano al Canal de Beagle, Tierra del Fuego, Argentina.* Unpublished Ph. D. thesis, Facultad de Ciencias Exactas y Naturales, Universidad de Buenos Aires, Buenos Aires.

Serrán, M., Centeno, N., Weiler, N. & Gómez Otero, J. (2008). Massive death of pinnipeds 1200 years ago: Taphonomic history of the ''Lobos site''

(Golfo Nuevo, Patagonia, Argentina). *Quaternary International 183*, 135-142.

Sielfeld, W., Venegas, C., Atalah A. & Torres, J. (1978). Prospección de otáridos en las costas de Magallanes. *Anales del Instituto de la Patagonia 9*, 157-170.

Smith, I. (2005). Retreat and Resilience: Fur Seals and Human Settlement in New Zealand. In Monks, G. (Ed). *The Exploitation and Cultural Importance of Sea Mammals, Proceedings of the 9 th Conference of International Council for Archaeozoology* (pp 6-18). Durham, UK: Oxbow Books.

Storå, J. (2001). Reading Bones. Stone Age Hunters and Seals in the Baltic. *Stockholm Studies in Archaeology 21*. Stockholm.

Storå, J. (2002). Neolithic Seal Exploitation on the Aland Island in the Baltic Sea on the Basis of Epiphyseal Fusion Data and Metric Studies. *International Journal of Osteoarchaeology 12*, 49-64.

In: Neotropical and Caribbean Aquatic Mammals ISBN: 978-1-63321-306-7
Editors: Muñoz, Götz and Ramos Roca

Chapter 7

HUMAN PREDATION ON PINNIPEDS IN THE BEAGLE CHANNEL

Atilio Francisco Zangrando*
CADIC – CONICET Bernardo Houssay Ushuaia,
Tierra del Fuego, Argentina

ABSTRACT

The model established for subsistence patterns of hunter-gatherer societies which inhabited the islands located southwards Tierra del Fuego argues that pinnipeds constituted the most important food resource. This general view is based on two fundamental points: (a) an increase in human carrying capacity due to energy importation produce through these marine mammals, (b) the independence of seal populations with respect to human activities, which guaranteed a sustained supply of animals to the human populations throughout the archaeological sequence. This model arises primarily from the evidence provided by the layer D of Túnel I archaeological site in the Beagle Channel, Tierra del Fuego Island. This chapter aims to make a reflection about this model. To do so, it brings together relevant information about pinnipeds representations in other archaeological sites of the Beagle Channel. In light of recent studies, it is noted that harvest strategies of seals would have been more variable than previously suggested, generating new questions about the relationship between humans and pinnipeds along occupational sequence of the Beagle Channel.

* E-mail: panchozan@yahoo.com.ar; afzangrando@cadic-conicet.gob.ar

INTRODUCTION

Pinnipeds have been important to the subsistence of maritime hunter-gatherers which inhabited the Beagle Channel region, Tierra del Fuego island, from 6,400 years BP to the 19th century. Moreover, for many colleagues prevails the idea that these resources were the basis of subsistence throughout the entire occupational sequence, providing an inexhaustible food for hunter-gatherer societies. Only in the nineteenth century and with the arrival of European and American sealers, the subsistence and behavioral patterns of these groups would have been seriously compromised (Orquera & Piana 1999a, 2009; Schiavini 1990, 1993). The investigations of Luis Orquera and Ernesto Piana (1986-87, 1999a, 2009) have generated important information and established the basis for our understanding of hunting and transportation technologies of pinnipeds in that region. However, the information about the butchery of carcasses is still poor and consequently we have a simplified view on the transport strategy of these resources. This chapter proposes a reflection on this general vision about the human-pinniped relation of the Beagle Channel.

In the last 35 years, different researches have compiled and categorized data in different analytical units, which were used to support their particular interpretations. In this chapter, the current state of knowledge regarding zooarchaeological representations of pinnipeds in the Beagle Channel is presented. Then this information is analyzed in light of competing interpretations of key researches. Although not all questions of earlier debates are addressed, two main discussions about the use of pinnipeds are proposed for the southern end of South America.

PINNIPEDS

Although seven species of pinnipeds (*Arctocephalus gazella*, *Arctocephalus tropicalis*, *Mirounga leonina*, *Leptonychotes weddellii*, *Lobodon carcinophga*, *Hydrurga leptonyx* and *Ommatophoca rossi*) can occasionally arrive to the coasts of Archipelago of Tierra del Fuego, only the South American Fur seals (*Arctocephalus australis*) and the Southern sea lions (*Otaria flavescens*) are permanently present (Bastida & Rodríguez 2003). There is considerable variability in body sizes in the two species, at both the inter-specific and intra-specific levels. According to records from Islas

Malvinas (Vaz Ferrerira 1979), the South American fur seal reaches the following dimensions: the pups weigh between 3.5 and 5.5 kg; 1.9 m to 159 kg in adult males; 1.4 m and 48.5 kg in adult females. Schiavini (1990, 1993) set up that the weight of adult males of *A. australis* ranges between 80.9 and 68.2 kilograms and females with an average of 40.6 kilograms. However, King (1983) notes that individuals of *A. australis* populations of Uruguay are much smaller than those of Malvinas Islands. The dimensions of Southern sea lions are larger (Vaz Ferreira 1979): the pups weigh between 10.2 and 14.8 kg; 2.56 m and 300-340 kg in adult males, 2 m and 144 kg in adult females; these data belong to both the Uruguayan coasts and the Islas Malvinas.

The majority of colonies of *A. australis* for mating and pupping (rookeries) are located on outer coasts and islands of the archipelago, away from the inner channels, while colonies of *O. flavescens* are distributed more homogeneously in the archipelago (Figure 1; Schiavini, 1990, 1993; Sielfeld et al. 1978; Falabella et al. 2009; Venegas et al. 2002). The population of *A. australis* is greater than *O. flavescens*. A survey conducted in the 1970s has indicated a population of 103,000 and 13,000 respectively, between Wellington islands and Cape Horn (Sielfeld et al. 1978). According to a census conducted more recently during the breeding season 2000-2001 by Venegas et al. (2002), this disparity is slightly inferior in the area between the Beagle Channel and Cape Horn. For this sector, populations of 4632 for *A. australis* and 912 for *O. flavescens* were computed. In the population of *A. australis* the 37% are females, 18% are pups and 17% are males (15% adults and 2% subadults). The population of *O. flavescens* has very similar proportions: 39% are females, 14% are pups and 14% are males (10% adults and 4% subadults) (Venegas et al. 2012).

Otariids have a wide distribution in Southern South America and can be found throughout the year as individuals moving in the sea or in permanent or temporary settlements on coasts and small islands. The age of sexual maturity is about 3 years old for female seals and 7 or 8 years old for male seals. Breeding occurs between November and December. Adult male seals defend territories throughout the breeding season, leaving the male seals which do not succeed in forming harems in the water around the breeding colonies. During the breeding season, the male seals of the harem do not feed, consuming fat reserves accumulated in pre-reproductive pelagic life. Reproductive groups disperse into January, and the female seals feed their pups during a period of 6 months to a year.

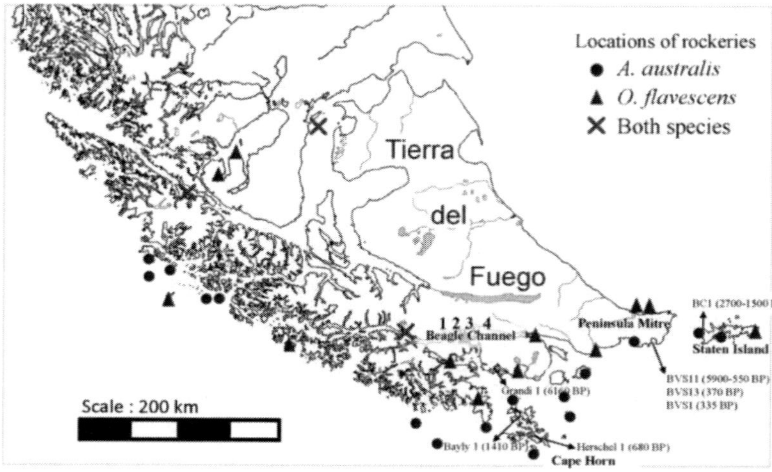

Figure 1. Locations of rockeries and of archaeological sites (1. Ajej I; 2. Lancha Packewaia; 3. Túnel I and Túnel VII; 4. Shamakush I).

Outside the breeding season, adult and young male seals spend more time in the water than on land, making long dispersions, but adult female seals must return continually to the breeding colonies to nurse their pups. In the breeding season, the female seals feed in areas near breeding colonies and the duration of foraging trips increases with time since the birth of the pups. Weaned pups remain feeding near the rockeries, but after a month, they move further (Schiavini, 1990, 1993; Bastida & Rodríguez 2003).

Otariids in the Archaeological Context

Table 1 summarizes the available information on the representations of otariids in archaeological sites in the central portion of the Beagle Channel.

The NISP value for pinniped remains from layer D of Túnel I site is approximately 55400 specimens (Orquera 2009 pers. comm.). Schiavini (1990: Table 29) established MNI values of 273 *A. australis* and 9 *O. flavescens* by maxillary and mandibular pieces. Calculations made by Orquera with post-cranial bones showed no significant differences in the abundance of both species (see Orquera & Piana 1999a: Table XV). Schiavini (1990: Figure 36) identified sex and age for 223 individuals of *A. australis*: 86.5% of the subjects were males.

Table 1. MNI values for otariids in archaeological contexts of the Beagle Channel

Site		Túnel I (Layer D)		Lancha Packewaia (Layer X)		Lancha Packewaia (Layers B, C and D)		Ajej I		Shamakush I		Túnel VII	
Reference		Schiavini 1990, Orquera pers. com.		Saxon 1979		Saxon 1979		Piana et al. 2008		Orquera & Piana 1996		Zangrando et al. 2010	
Age (^{14}C yr BP)		6200-4500		4020		1500-280		1400		1000		100	
NISP		55400		1429		2691		1037		169		8953	
A. australis (MNI)		273		16		66		11		5		40	
Age categories	Pup	18	8.1%	4	25%	5	7.6%	0	0%	3	60%	9	22.5%
	Subadult	134	60.1%	8	50%	46	69.7%	11	100%	1	20%	15	37.5%
	Adult	71	31.8%	4	25%	15	22.7%	0	0%	1	20%	16	40%
Sex	Male	193	86.5%	1	100%	15	83.3%	2	18.2%	no data		8	47%
	Female	30	13.5%	0	0%	3	16.7%	9	81.%	no data		9	53%
O. flavescens (MNI)		9		2		4		0		0		3	
Age categories	Pup	no data		0	0%	0	0%	0	0%	0	0%	0	0%
	Subadult	no data		2	100%	3	75.0%	0	0%	0	0%	1	33.3%
	Adult	no data		0	0%	1	25.0%	0	0%	0	0%	2	66.7%
Sex	Male	?		no data		no data		0	0%	0	0%	1	50,0%
	Female	Dominate		no data		no data		0	0%	0	0%	1	50.%

All ages are represented in layer D of Túnel I, but the 69% of the males were under 8 years (non reproductive males), while in females less than 4 years were represented by 37% (Schiavini 1990). Schiavini (1990) also established the months of death for different specimens: the obtained distribution was few deaths in February, a considerably higher frequency between April and July, a progressive decrease until November, and no deaths recorded in December and January. In conclusion, 90% of pinnipeds represented in the layer D would have died in the seven months between March and September (Schiavini 1990: Figure 42).

Saxon (1979) initially reported 1761 identified specimens of pinniped from Lancha Packewaia site; that value was modified years later by 4120 specimens (Orquera & Piana 1993-1994: Table 16). Saxon classified the remains of pinniped into three age classes and without reporting sex categories. Subsequently Schiavini (1990) made new determinations on this assemblage studying maxillary and mandibular pieces. Saxon (1979) identified 82 individuals for layers B, C, D and X of Lancha Packewaia, but Schiavini found only 36 maxillary and 46 mandibles and not all of these fragments were preserved canines. Therefore, Schiavini could only make determinations of 16 males and 3 females from those layers.

Orquera classified the remains of pinniped from Shamakush I into three classes, identifying one large pinniped, one medium-small individual and three small individuals (Orquera & Piana 1996). Schiavini (1990: 100) noted that all maxillary and mandibular pieces (three in total) correspond to an *A. australis*.

In Ajej I, Vázquez identified 1037 remains corresponding to *A. australis* and determined 11 individuals. Schiavini determined that about 80% of individuals are females and all juveniles, mostly between three and seven years old (Piana et al. 2008).

Maxillaries from Túnel VII site allowed determining a MNI of 7 individuals: a male subadult of *O. flavescens*, an adult male of *A. australis*, three juveniles, an adult female and a pup (Estévez et al. 1995). These authors also analized anatomical representations and bone modifications on a sample of 693 specimens. Then, Zangrando (2009a) continued with the analysis of this site and identified a higher NISP and MNI of 40 *A. australis* and 3 *O. flavescens*. From sizes and fused states of long bones, Zangrando differed between adult individuals (completely fused bones), subadults (not fused or partially fused bones) and pups (not fused and very small size bones). Sex determination was performed by coxal bones.

Thus Table 1 summarizes the information published over the past 35 years regarding the representations of otariids in the archaeological record of the north coast of the Beagle Channel. This table shows a large predominance of *A. australis* and a predominance of subadults of this species in the archaeological context. However, while pups are underrepresented in many sites, it is interesting to note that this category also reaches 25% in some sites, separating the exceptional case of Shamakush I. Regarding sex, an interesting variation is also recorded in *A. australis*, in which females can reach up to 82% (*v.g.* Ajej I).

DISCUSSION

A. Australis … How Intensive?

Schiavini (1990, 1993) has argued that human predation in the central section of the Beagle Channel would not have affected pinniped populations primarily by lack of access to rockeries. Zooarchaeological studies based on determination of sex, age and season of death in maxillaries, all recovered in the layer D of Túnel I (6,200-4,300 years BP), showed that capture patterns in the central portion of Beagle Channel occurred predominantly in the water rather than in rockeries: the captures were mainly males (83%) and concentrated between Autumn and Spring (Schiavini 1990, 1993). A capture focused on this age-sex category would have not generated an overexploitation and would have allowed pinnipeds to be the main resource to human subsistence throughout the occupational sequence without producing an impact on the population structure or their abundance in the environment (Schiavini 1990, 1993; Orquera & Piana 1999a). However, recent studies have revealed long-term variations in the representations of this resource along the archaeological sequence of the Beagle Channel (Zangrando 2009a, 2009b; Zangrando et al. 2010; Tivoli & Zangrando 2011). During the first two millenniums of the sequence, the zooarchaeological assemblages are characterized by high inter-taxonomic dominance of pinnipeds, while indices representing other vertebrates (e.g. guanacos, birds and fish) remain lower. From 5,000 years BP to later times, a decrease in the relative importance of pinnipeds and a diversification in subsistence patterns are recorded, which was marked by a significant increase in the consumption of low-ranked preys (birds and fish) during the Late Holocene (Zangrando 2009a). Notwithstanding the influence of paleoenvironmental factors (Saporiti et al. 2013), it was suggested that human predation pressure seems to have played a significant role in the availability of *A. australis* in the Archipelago of Tierra del Fuego through the Holocene (Zangrando et al. 2010; Zangrando et al. 2013).

If captures were focused on subadult males of *A. australis*, which would not have affected the demographic structure of this resource, why does the representation of pinnipeds decrease over time? Schiavini (1990) based his arguments on the evidence of layer D of Túnel I, which are factually and methodologically forceful. However, the information from that layer indicates a type of human-prey relation for the central part of the Beagle Channel and for early stages of the occupational sequence. While it is true that remains of subadult males of *A. australis* are conspicuous in the regional context, the archaeofaunal record also shows variations through different sites ranging from the intensive use to sporadic exploitation of females in archaeological sites of the late Holocene; the higher representations of females in Ajej I and Túnel VII are suggestive in this regard (Table 1).

Moreover, the *A. australis*-human relation was not limited to the Beagle Channel sector and the relation inferred from the Beagle Channel record could have been different to that developed in other areas of the archipelago. Exploring how capture patterns of *A. australis* were developed in locations near the rockeries is a requirement to understand how human predation could affect the populations of that species during the Holocene. Archaeological studies developed in the Cape Horn archipelago showed occupations between 1,400 and 680 years BP (Legoupil 1993-94). Studies conducted in Crossley Bay I site, Staten Island, proved that this island was visited between 2,700 and 1500 years BP (Horwitz & Weissel 2011). In Valentin Bay, at the south coast of Peninsula Mitre, occupations between 5,000 years BP and historical moments were also recorded (Vázquez et al. 2011). The archaeological record of all these locations include remains of pinnipeds and nowadays rockeries of *A. australis* are common nearby those islands and coastal places.

O. Flavescens... Why So Little?

O. flavescens is generally poorly represented in archaeological sites of the Patagonian archipelago (Muñoz 2011), being Punta Baja the only exception with an archaeofaunal record dominated by female individuals (Legoupil 1989). The underrepresentation of adult males of *O. flavescens* is particularly noticeable, and the immediate question thus becomes why the larger prey that the Otariidae Family offers is weakly represented in the archaeological record?

It was noted that this disparity between the two species represented in the archaeological record matches with the natural numerical relation existing in the environment (Orquera & Piana 1999a: 69). Though it is not possible to assess to what extent the current abundance and distribution of pinnipeds in the Archipelago of Tierra del Fuego are still distorted by the overexploitation of europeans and americans during the nineteenth and early twentieth centuries, an exploration with the current ecological parameters is equally useful. The information presented by Venegas *et al.* (2002) from the quantification of pinnipeds during the reproductive season 2000-2001 for surrounding areas of the Beagle Channel (see synthesis above) shows that the ratio is 5 *A. australis* by 1 *O. flavescens*, including both sexes and all age groups. Although this disparity coincides with that observed in the archaeological record, the numerical relation in the modern populations of pinnipeds is significantly lower to that inferred from the archaeological context. Moreover, if we consider the distribution of pinnipeds presented in Figure 1, the locations of archaeological sites show greater spatial relationship with colonies and rockeries of *O. flavescens* and it thus seems this species was more accessible to humans. Modern environmental parameters do not appear to constitute sufficient evidence to explain the underrepresentation of *O. flavescens* in the archaeological record.

As it has been shown in empirical ethnographic and experimental studies (e.g. Kelly 1995; Winterhalder 1981), larger prey are usually ranked higher than smaller animals. However, some resources (e.g. whales or some pinnipeds) have excessive handling costs that depress their rank relative to overall body size (e.g. Byers & Ugan 2005; Jones 2004). Abundant zooarchaeological, technological and ethnographic information from the Archipelago of Tierra del Fuego suggest that the captures of pinnipeds occurred mainly in the sea using canoes and harpoons (Orquera & Piana 1999a, 1999b). This leads to think that aquatic transport of hunted prey could have been a key factor in the decisions of the foragers.

Orquera and Piana (1999b: 260-262) analyzed ethnographic information related to canoes of hunter-gatherer-fishers from the archipelago and estimated that this watercraft could carry as much as 300 to 400 kg. Transporting an adult male of *O. flavescens* would have clearly over-passed that capacity with the serious risks that it would have involved. In contrast, the transport of complete specimens of *A. australis* or *O. flavescens* females would have been achieved without much difficulty. Assuming that the dimensions of prehistoric canoes were similar to those seen in historical times, the adult males of *O. flavescens* would not have been presented as an optimal choice when hunting occurred at sea. Martinoli (2012) suggests that this prey has been opportunistically exploited in the coasts of the Beagle Channel.

The underrepresentation of larger pinnipeds may also respond to other factors. Martinoli (2012) has developed a model to analyze the transport and processing of both species which suggests that adult male specimens of *O. flavescens* were necessary butchered prior to transport, while complete carcass of pups and females of this species and both sexes and all age groups of *A. australis* could have been transported to base camps or places of consumption. This author has observed that most of anatomical portions of *A. australis* are represented in archaeological sites of the Beagle channel, while very partial presence of units of larger *O. flavescens* is normally recorded. Therefore, different butchering pattern of adult males of *O. flavescens* over other pinnipeds would have also produced a differential contribution of bones to the archaeological record.

CONCLUSION

The different investigations and the zooarchaeological evidence available are inconclusive about the intensity of hunter-gatherer predation on *A. australis* in the Beagle Channel and nearby areas. Improving understanding of pinniped use will involve more systematic studies, specially from sites located in outer parts of the archipelago (v.g. Staten Island; Archipelago of Cape Horn, etc.).

Since body size plays an important role in the decisions of carcass processing (e.g. Lyman 1991), adult males of *O. flavescens* may involve different strategies that other otariids (Martinoli 2012). Such possibility should be considered along with analyses of butchery patterns and transportation capabilities.

ACKNOWLEDGMENT

I am grateful to Sebastián Muñoz, Christopher Markus Götz and Elizabeth Ramos Roca for inviting me to participate in this book. Luis Orquera, Marcela Liljesthrom and Angélica Tivoli made very useful comments on previous versions of this manuscript. The suggestions of an anonymus reviewer were also appreciated. This research was supported by PIP 0395/10 (CONICET),

PICT-2010-1322 (FONCYT) and the "Marine Ventures" Project (Research Counsil of Norway - no 208828). Excavations and research in Túnel VII site were conducted by J. Estévez Escalera and A. Mitja Vila.

REFERENCES

Bastida, R. & Rodriguez, D. (2003). *Mamíferos Marinos. Patagonia-Antartida.* Buenos Aires, Argentina: Vazquez Mazzini Editores.

Binford, L. (1981). *Bones: Ancient Men and Modern Myths.* Nueva York, USA, Academic Press.

Byers, D. A. & Ugan, A. (2005). Should we expect large game specialization in the Late Pleistocene? An optimal foraging perspective on early Paleoindian prey choice. *Journal of Archaeological Science 32*, 1624–1640.

Estévez, J., Juan-Muns, N., Martínez J., Piqué R. & Schiavini, A. (1995). Zooarqueología y Antracología: estrategias de aprovisionamiento de los recursos animales y vegetales en Túnel VII. In Estévez, J. & A. Vila (Eds.). *Treballs d'Etnoarqueologìa 1, Encuentros en los conchales fueguinos* (pp. 143-238). Barcelona, España: CSIC / Universidad Autónoma de Barcelona.

Horwitz V. & Weissel, M. (2011). Arqueología de Isla de los Estados: la frontera de la abundancia. In Zangrando A. F., M. Vázquez & A. Tessone (Eds). *Los cazadores-recolectores del extremo oriental fueguino. Arqueología de Península Mitre e Isla de los Estados* (pp. 271-286). Buenos Aires; Argentina: Sociedad Argentina de Antropología.

Jones, E. L. (2004). Dietary evenness, prey choice, and human-environment interactions. *Journal of Archaeological Science 31*, 307–317.

Kelly, R. L. (1995). *The Foraging Spectrum. Diversity in Hunter-Gatherer Lifeways.* Washington, USA: Smithsonian Institution Press.

King, J., (1983). *Seals of the world. Londres.* Oxford, UK: Oxford University Press.

Falabella, V., Campagna C. & Croxall J. (2009). *Atlas del Mar Patagónico. Especies y espacios.* Buenos Aires, Argentina: Wildlife Conservation Society and Bird Life International.

Legoupil, D. (1989). *Ethno-archeólogie dans les archipels de Patagonie: les nomades marines de Punta Baja.* París, France: Editions Resherche sur les Civisations.

Legoupil D. (1993-94). El archipiélago del cabo de Hornos y la costa sur de la isla Navarino: poblamiento y modelos económicos. *Anales del Instituto de la Patagonia, Serie Ciencias Humanas 22,* 101-121.

Lyman, R. L. (1991). *Prehistory of the Oregon Coast: The effects of Excavations Strategies and Assemblage Size on Archaeological Inquiry.* San Diego, USA, Academic Press.

Martinoli M. P. (2012). *Modalidades de explotación de pinnípedos en las ocupaciones canoeras tempranas del canal Beagle: el sitio Imiwaia I (Tierra del Fuego, Argentina).* Unpublished PhD dissertation. Facultad de Filosofía y Letras, Universidad de Buenos Aires.

Muñoz, A. S. (2011). Pinniped Zooarchaeological Studies in Southern Patagonia: Current Issues and Future Research Agenda. In Bicho, N., J. Haws & L. Davis (Eds.). *Trekking the Shore. Changing Coastlines and the Antiquity of Coastal Settlement* (pp. 305-331). Heilderbergh, Germany, Springer.

Orquera, L. A. & Piana, E. L. (1986-87). Composición topológica y datos tecnomorfológicos y tecnofuncionales de los distintos conjuntos arqueológicos del sitio Túnel I (Tierra del Fuego). *Relaciones de la Sociedad Argentina de Antropología XVII*, 201-239.

Orquera, L. A. & Piana, E. L. (1993-94). Lancha Packewaia: actualización y rectificaciones. *Relaciones de la Sociedad Argentina de Antropología, XIX*, 325-362.

Orquera, L. A. & Piana, E. L. (1996). El sitio Shamakush I (Tierra del Fuego, República Argentina). *Relaciones de la Sociedad Argentina de Antropología, XXI,* 215-265.

Orquera, L. A. & Piana, E. L. (1999a). *Arqueología de la región del canal Beagle (Tierra del Fuego, República Argentina).* Buenos Aires, Argentina: Sociedad Argentina de Antropología.

Orquera, L. A. & Piana, E. L. (1999b). *La vida material y social de los Yámana*. Buenos Aires, Argentina: Editorial Universitaria de Buenos Aires.

Orquera, L. A. & Piana, E. L. (2009). Sea nomads of the Beagle Channel in southernmost South America: Over six thousand years of coastal adaptation and stability. *Journal of Island and Coastal Archaeology 4*, 1–21.

Piana, E. L., Vázquez M. & Álvarez, M. (2008). Nuevos resultados del estudio del sitio Ajej I: un aporte a la variabilidad de estrategias de los canoeros fueguinos. *Runa 29*, 101-121.

Saporiti F., Bala L. O., Crespo E. A., Gómez-Otero J., Zangrando A. F., Aguilar A. & Cardona, L. (2013). Changing patterns of marine resource exploitation by hunter-gatherers throughout the late Holocene of Argentina are uncorrelated to climatic variability. *Quaternary International 299*, 108–115.

Saxon, E. C. (1979). Natural Prehistory: the archaeology of Fuego-Patagonian ecology. *Quaternaria XXI*, 329-356.

Schiavini, A. M. (1990). *Estudio de la relación entre el hombre y los pinnípedos en el proceso adaptativo humano del canal Beagle (Tierra del Fuego, Argentina).* Unpublished PhD dissertation. Facultad de Ciencias Exactas y Naturales, Universidad de Buenos Aires, Buenos Aires.

Schiavini, A. M. (1993). Los lobos marinos como recurso para cazadores-recolectores marinos: el caso de Tierra del Fuego. *Latin American Antiquity,* 4: 346-366.

Sielfeld, W., Venegas, C., Atalah, A. & Torres, J. (1978). Prospección de otáridos en las costas de Magallanes. *Anales del Instituto de la Patagonia IX,* 157-169.

Tivoli A. & Zangrando A. F. (2011). Subsistence variations and landscape use among maritime hunter-gatherers. A zooarchaeological analysis from the Beagle Channel (Tierra del Fuego, Argentina). *Journal of Archaeological Science 38*, 1148-156.

Vaz Ferreira, R. (1979). South American sea lion. *FAO Fisheries Series 5(2),* 9-12.

Vázquez M., Zangrando A. F., Tessone, A. & Ceraso, A. (2011). Investigaciones arqueológicas recientes en la costa suroriental de Península Mitre. Los cazadores recolectores del extremo oriental fueguino. In Zangrando A.F., M.Vázquez & A. Tessone (Eds). *Los cazadores-recolectores del extremo oriental fueguino. Arqueología de Península Mitre e Isla de los Estados* (pp. 203-230). Buenos Aires; Argentina: Sociedad Argentina de Antropología.

Venegas, C., Gibbons, J., Aguayo, A., Sielfel, W., Acevedo, J., Amado, N., Capella, J., Guzmán, G., & Valenzuela, C. (2002). Distribución y abundancia de lobos marinos (Pinnipedia: Otariidae) en la Región de Magallanes, Chile. *Anales del Instituto de la Patagonia 30*, 67-82.

Winterhalder, B. (1981). Foraging strategies in the boreal environment: An analysis of Cree hunting and gathering. In Winterhalder B. & E. A. Smith (Eds.). *Hunter-gatherer Foraging Strategies. Ethnographic and Archaeological Analyses* (pp. 66-98). Chicago, USA: The Chicago of University Press.

Zangrando A. F. (2009a). *Historia evolutiva y subsistencia de cazadores-recolectores marítimos de Tierra del Fuego. Colección Tesis de Doctorado.* Buenos Aires, Argentina: Sociedad Argentina de Antropología.

Zangrando A. F. (2009b). Is fishing intensification a direct route to hunter-gatherer complexity? A case study from the Beagle Channel region (Tierra del Fuego, southern South America). *World Archaeology 41 (4)*, 589-608.

Zangrando A. F, Orquera L. A. & Piana E. L. (2010). Diversificación e intensificación de recursos animales en la secuencia arqueológica del canal Beagle (Tierra del Fuego, Argentina).). In Gutiérrez, M., M. De Nigris, P. Fernández, M. Giardina, A. Gil, A. Izeta, G. Neme & H. Yacobaccio (Eds.). *Zooarqueología a principios del siglo XXI: aportes teóricos, metodológicos y casos de estudio* (pp. 359-370). Buenos Aires, Argentina: Ediciones del Espinillo.

Zangrando A. F., Panarello H. & Piana E. (2013). Zooarchaeological and Stable Isotopic Assessments on Pinniped–Human Relations in the Beagle Channel (Tierra del Fuego, Southern South America). *International Journal of Osteoarchaeology* (In press).

In: Neotropical and Caribbean Aquatic Mammals ISBN: 978-1-63321-306-7
Editors: Muñoz, Götz and Ramos Roca

Chapter 8

MARINE MAMMALS IN SOUTHERN SOUTH AMERICA: BIOLOGY, CURRENT STATUS AND CONSERVATION

Juan Ignacio Túnez[1,*] and Lida Elena Pimper[2,†]
[1]Grupo de Estudios en Ecología de Mamíferos,
Departamento de Ciencias Básicas,
Universidad Nacional de Luján y CONICET
[2]Museo Acatushún de Aves y Mamíferos Marinos Australes,
Tierra del Fuego, Argentina

ABSTRACT

Marine ecosystems of southern South America are very rich in species of marine mammals, with 10 pinniped and 46 cetacean species reported up to date. This chapter starts with a short description on the recorded presence and conservation status of these marine mammal species in the area. Then, in separate sections, each one corresponding to a different species, the reader will find the state of art on studies about biology, population numbers and trends, mayor threats, conservation genetics and population structure of six of the most emblematic species, extensively studied in the last decades, that inhabit southern South America coasts and waters.

* E-mail: nacho_tunez@yahoo.com.ar
† E-mail: liditas@gmail.com

Introduction

Marine ecosystems of southern South America are very rich in species of marine mammals. Fifty-six species have been recorded up to date in its coasts and waters (Table 1), comprising 10 pinnipeds and 46 cetaceans. Information about current conservation status and population trends of almost 50% of these species is not available, while others have been extensively studied in the last decades. In this chapter, we provide an updated report on the biology, population numbers and trends, mayor threats and conservation genetics and population structure of six of the most emblematic marine mammal species currently found in southern South America: the South American sea lion (*Otaria flavescens*), the South American fur seal (*Arctocephalus australis*) the southern elephant seal (*Mirounga leonina*), the southern right whale (*Eubalaena australis*), Commerson's dolphin (*Cephalorhynchus commersonii commersonii*) and Franciscana dolphin (*Pontoporia blainvillei*).

South American Sea Lion, *Otaria Flavescens*

Geographic Range, Population Numbers and Trends

South American sea lions are distributed along approximately 10,000 km of the coast of South America, from Recifes das Torres in southern Brazil to Cape Horn in the Atlantic coast, and from there along the Pacific coast to Zorritos, in northern Peru (Riedman 1990). They also occur in Malvinas (Falkland) Islands (King 1983). On the Atlantic coast, the species present three areas with breeding activity, the Uruguayan coasts, north-central Patagonia and southern Tierra del Fuego, Argentina (Túnez et al. 2008a). On the Pacific, distribution of breeding colonies is almost continuous (Sielfeld 1999). Total population was estimated in 250,000 individuals, with about 12,000 individuals in Uruguay, no less than 100,000 in continental Argentina, 6,000 in Malvinas (Falkland) Islands, 90,000-100,000 in Chile and 60,000 in Peru (Campagna 2008). Population numbers are declining in Peru, Malvinas (Falkland) Islands and Uruguay (Arias-Schreiber & Rivas 1998, Thompson et al. 2005; Páez 2005). In central Chile, population numbers were almost stable between 1985 and 2007 (Sepúlveda et al. 2011). Pup numbers in north-central Patagonia, Argentina, are increasing at an annual rate of 5.7% (Dans et al. 2004). Estimates for southern Patagonia show a decrease to only 14.5% of the population size reported in the late 1940 (Schiavini et al. 2004).

Table 1. Marine mammal species recorded in the southern cone of South America and its current conservation status

Specific name	Common name	Recorded presence	Conservation status
Pinnipeds			
Otariidae			
Arctocephalus gazella	Antarctic fur seal	Erratic	Least concern
Arctocephalus tropicalis	Sub Antarctic fur seal	Erratic	Least concern
Arctocephalus australis	South America fur seal	Atlantic and Pacific	Least concern
Arctocephalus philippii philippii	Juan Fernandez fur seal	Juan Fernandez Islands	Near Threatened
Otaria flavescens	South American sea lion	Atlantic and Pacific	Least concern
Phocidae			
Mirounga leonina	Southern elephant seal	Atlantic and Pacific	Least concern
Leptonychotes weddellii	Weddell seal	Erratic	Least concern
Ommatophoca rossii	Ross seal	Erratic	Least concern
Lobodon carcinophaga	Crabeater seal	Erratic	Least concern
Hydrurga leptonyx	Leopard seal	Erratic	Least concern
Cetacea			
Mysticeti (baleen whales)			
Balaenidae (right whales)			
Eubalaena australis	Southern right whale	Atlantic and Pacific	Least concern
Caperea marginata	Pygmy right whale	Atlantic and Pacific	Data Deficient
Balaenopteridae (rorquals)			
Megaptera novaeangliae	Humpback whale	Atlantic and Pacific	Least concern
Balaenoptera acutorostrata	Dwarf minke whale	Atlantic and Pacific	Least concern
Balaenoptera bonaerensis	Antarctic minke whale	Erratic	Data Deficient
Balaenoptera edeni	Bryde's whale	Atlantic and Pacific	Data Deficient

Table 1. (Continued)

Specific name	Common name	Recorded presence	Conservation status
Balaenoptera borealis	Northern sei whale	Atlantic and Pacific	Endangered
Balaenoptera physalus	Fin whale	Atlantic and Pacific	Endangered
Balaenoptera musculus	Blue whale	Atlantic and Pacific	Endangered
Odontoceti (toothed whales, dolphins and porpoises)			
Physeteridae			
Physeter macrocephalus	Sperm whale	Atlantic and Pacific	Vulnerable
Kogiidae			
Kogia breviceps	Pygmy sperm whale	Atlantic and Pacific	Data Deficient
Kogia sima	Dwarf sperm whale	Atlantic and Pacific	Data Deficient
Ziphiidae (beaked whales)			
Ziphius cavirostris	Cuvier's beaked whale	Atlantic and Pacific	Least concern
Berardius arnuxii	Arnoux' beaked whale	Atlantic and Pacific	Data Deficient
Tasmacetus shepherdi	Shepherd's beaked whale	Atlantic and Pacific	Data Deficient
Hyperoodon planifrons	Southern bottlenose whale	Atlantic and Pacific	Least concern
Mesoplodon hectori	Hector's beaked whale	Atlantic and Pacific	Data Deficient
Mesoplodon grayi	Gray's beaked whale	Atlantic and Pacific	Data Deficient
Mesoplodon layardii	Strap-toothed beaked whale	Atlantic and Pacific	Data Deficient
Mesoplodon densirostris	Blainville's beaked whale	Atlantic	Data Deficient
Mesoplodon peruvianus	Pygmy beaked whale	Pacific	Data Deficient
Mesoplodon traversii	Spade-toothed whale	Pacific	Data Deficient
Pontoporiidae			
Pontoporia blainvillei	Franciscana dolphin	Atlantic	Vulnerable
Delphinidae			
Cephalorhynchus commersonii commersonii	Commerson's dolphin	Atlantic	Data Deficient

Specific name	Common name	Recorded presence	Conservation status
Cephalorhynchus eutropia	Chilean dolphin	Pacific	Near Threatened
Steno bredanensis	Rough-toothed dolphin	Atlantic and Pacific	Least Concerned
Tursiops truncatus	Common bottlenose dolphin	Atlantic and Pacific	Least Concerned
Stenella attenuata	Pantropical spotted dolphin	Atlantic and Pacific	Least Concerned
Stenella frontalis	Atlantic spotted dolphin	Atlantic	Data Deficient
Stenella clymene	Clymene dolphin	Atlantic	Data Deficient
Stenella coeruleoalba	Striped dolphin	Atlantic	Least Concerned
Delphinus delphis	Short-beaked common dolphin	Atlantic and Pacific	Least Concerned
Delphinus capensis	Long-beaked common dolphin	Atlantic	Data Deficient
Lagenodelphis hosei	Fraser's dolphin	Atlantic	Least Concerned
Lagenorhynchus obscurus	Dusky dolphin	Atlantic and Pacific	Data Deficient
Lagenorhynchus australis	Peale's dolphin	Atlantic and Pacific	Data Deficient
Lagenorhynchus cruciger	Hourglass dolphin	Atlantic and Pacific	Least Concerned
Lissodelphis peronii	Southern right-whale dolphin	Atlantic and Pacific	Data Deficient
Grampus griseus	Risso's dolphin, gray grampus	Atlantic and Pacific	Least Concerned
Feresa attenuata	Pygmy killer whale	Atlantic	Data Deficient
Pseudorca crassidens	False killer whale	Atlantic and Pacific	Data Deficient
Orcinus orca	Killer whale	Atlantic and Pacific	Data Deficient
Globicephala melas	Long-finned pilot whale	Atlantic and Pacific	Data Deficient
Globicephala macrorhynchus	Short-finned pilot whale	Atlantic and Pacific	Data Deficient
Phocoena spinipinnis	Burmeister's porpoise	Atlantic and Pacific	Data Deficient
Phocoena dioptrica	Spectacled porpoise	Atlantic and Pacific	Data Deficient

Note: Family names used correspond to those of the first official Society for Marine Mammalogy list of marine mammal species (Committee on Taxonomy, 2011). Data sources: Redford & Eisenberg (1992), Perrin et al. (2009), IUCN (2013).

Conservation Status

The species is currently categorized as "least concerned" by the IUCN (International Union for Conservation of Nature) and is not included in any CITES appendix (the Convention on International Trade in Endangered Species of Wild Fauna and Flora).

Habitat and Ecology

South American sea lions are strongly sexually dimorphic (Cappozzo 2002). Adult males reach 2.83 m in length and 354 kg in weight, while females reach 1.92 m and 150 kg. At birth, pups are 72-89 cm long and weight between 10 and 17 kg (Cappozzo et al. 1991). Female sea lions attain its sexual maturity at 2-4 years, while males reach it at 5-7 years. Longevity is considered to be 18-20 years (Crespo 1988).

Sea lions prefer slightly slope coast to reproduce (Túnez et al. 2008a). The start of the breeding season varies by location and latitude. At most breeding colonies, males arrive in mid-December and gravid females between mid-December and early-January (Campagna & LeBoeuf 1988). Females give birth 2-3 days after their arrival at the colonies, with a peak of birthing in mid-January (Campagna 1985). Estrous occurs 6-8 days after parturition (Cappozzo 1995) and most copulas between mid-Decembre and mid-February (Campagna & Le Boeuf 1988).

South American sea lions are a highly polygynous species. During the breeding season, a well defined social organization is observed. Social groups are composed by a dominant male and 4 to 10 adult females, although some solitary couples are found. This variation in female numbers depends on the various strategies employed by males and females during the breeding season that are related to colony substrate, thermoregulatory requirements or avoiding of male harassment (Cassini 1999, 2000; Cappozzo et al. 2008).

Sea lions feed in shallow waters over the continental shelf (Campagna et al. 2001). Mean depth of lactating female foraging dives is 61 m and mean duration is 3 minutes, while males forage at a mean depth of 62.5 m (Müller et al. 2006). The species is considered an opportunistic feeder taking a variety of prey that include fishes, cephalopods, crustaceans, gastropods, polychetes, sponges, and tunicates (Koen Alonso et al. 2000; Szteren et al. 2004).

Major Threats

South American sea lions were hunted by native people for thousands of years (see Chapters 12 and 13) and have been taken by Europeans as early as the 16th century for food, oil and hides (Rodriguez & Bastida 1998).

Significant commercial harvests occurred in several countries and most populations were drastically reduced to <10% of their original numbers in the last several hundred years (Crespo & Pedraza 1991; Reyes et al. 1999; Schiavini et al. 2004). As a consequence, sea lions colonies disappeared from part of their range.

The species interacts regularly with fisheries (Sepúlveda et al. 2007). During the 1990s, mortality caused by fisheries in the Patagonian shelf varied from 150-600 sea lions per year (Dans et al. 2003). Intensive trawl fishing has been implicated in a severe decline of sea lions in Malvinas (Falkland) Islands (Thompson et al. 1998). Competition for fish between sea lions and artisan fisheries has been described in most of their range. Sea lions, according to fishermen, prey on fish caught in their fishing gear, often causing damage. However, no significant effects over artisan fisheries catches were found (Sepúlveda et al. 2007). In Chile, sea lions also interact with salmon farms.

Conservation Genetics and Population Structure

In the last decades, several studies analyzed the genetic structure and existence of conservation units in the species. Szapkievich et al. (1999) analyze nine serum protein loci in two colonies located in Uruguay and Northern Patagonia. They do not found significant genetic differences between colonies, concluding that colonies belong to the same population in which gene flow currently occurs. More recently, Túnez et al. (2007) analyze mitochondrial DNA (mtDNA) cytb sequences in six colonies from north-central Patagonia and Uruguay and compare their results with previously published sequences from a Peruvian colony (Wynen et al. 2001). They found no shared haplotypes between Atlantic and Peruvian colonies, suggesting the existence of different Evolutionary Significant Units (ESUs) in each ocean basin. Uruguayan and Patagonian colonies were also genetically different suggesting the existence of two Conservation Units (CUs) in the Atlantic. Feijoo et al. (2011) found the same result using mtDNA control region sequences. However, population structure was not detected when microsatellite markers were used. The authors interpret these differences between molecular markers as a consequence of female phylopatry and high male dispersion. Three years later of their first work, Túnez et al. (2010), using mtDNA control region sequences, found that north-central Patagonian colonies would have suffered a demographic contraction during a glacial period occurred 64,000 years before present (YBP), followed by a demographic expansion of the southernmost colonies when the glaciers retracted. No genetic data for Chilean populations are currently available.

South American Fur Seal, *Arctocephalus Australis*

Geographic Range, Population Numbers and Trends

The South American fur seal present almost the same geographic range as *O. flavescens*. It is distributed from Rio Grande do Sul in southern Brazil, to Cape Horn, and from there to Paracas Peninsula, in northern Peru (Riedman 1990). Mayor breeding colonies are not homogeneously distributed along the species distribution. They are found at three areas in the northern and southern extremities of the species range: the Uruguayan islands, the southern coast of Chile and Staten Island; and the northern coasts of Chile and Peru (Túnez et al. 2008b). Some colonies were also recorded during the breeding season in Malvinas (Falkland) Islands. The Uruguayan population of the species is the most abundant, with 200,000-250,000 individuals (Vaz-Ferreira 1982). The Argentinean population was estimated in 20,000 individuals, located in 17 island colonies southern to 43°S with low reproductive activity (Crespo et al. 1999). Chilean and Peruvian populations were estimated in 105,000 and 11,000 individuals, respectively (Arias-Schreiber & Rivas 1998, Sielfeld 1999). Data on population trends are scarce. Population of coastal Patagonia is increasing at about 8% per year (Crespo et al. 1999). El Niño events have dramatic effects on the Peruvian population, which declined 72% as a result of low food availability during the severe el El Niño in 1997-1998 (Oliveira et al. 2006).

Conservation Status

The species is currently categorized as “least concerned” by the IUCN and listed in CITES Appendix II. It is currently protected and managed by national and local laws along most of its geographical range.

Habitat and Ecology

South American fur seals are sexually dimorphic. Adult males reach 1.9 m in length and 120-160 kg in weight; while females are about 1.4 m long and weigh 40-50 kg. At birth, pups are 60-65 cm long and weight 3.5-5.5 kg (Cappozzo 1995). Female fur seals attain its sexual maturity at 2-3 years, while males reach it at 5-7 years (Vaz-Ferreira 1982). Longevity is considered to be 15.5 years for males and 16.5 for females (Batallés et al. 1990).

Colonies are generally found in rocky coasts. Breeding season starts in late-October, when adult males begin to settle in the colonies, and last to mid-January. Timing of breeding may differ between colonies according to latitude (H. Pavéz, unpublished). Females give birth shortly after they settle in the colonies. The peak of birthing occurs between last-November and early December. Estrous occur 5-8 days later (Cappozzo 1995) and most of the copulas between the second and third week of December (Vaz-Ferreira & Ponce de León 1987).

South American fur seals show a polygynous mating system. As in *O. flavescens*, a well defined social organization is observed during the breeding season. Social groups are composed by a dominant male and 1 to 13 females (Vaz-Ferreira 1982).

Distribution of fur seals at sea is poorly known. The species is thought to forage off the continental shelf (Naya et al. 2002, Thompson et al. 2003). Mean depth of female foraging dives is about 20 m, with a mean duration of 1 minute (York et al. 1998). Demersal and pelagic fishes are main preys (Naya et al. 2002). Cephalopods, lamellibranches, gastropods and crustaceans are also taken.

Mayor Threats

The species has been harvested by aborigines in Tierra del Fuego, Argentina, Uruguay and Peru, which used it as a source of meat and leather (see Chapters 7 and 8 in this volume). Since the arrival of Europeans to South America, fur seals have been heavily exploited due to the high quality of its fur. As a consequence, *A. australis* populations were practically decimated and most of its colonies were extinct (Cabrera & Yepes 1940, Ximénez & Langguth 2002). A managed harvest of adult males continued in Uruguay until 1991 (Vaz-Ferreira & Ponce de Leon 1987). In southern Chile, fur seals were illegally exploited as a source of free bait for the king crab fishery (Campagna 2008).

The species does not usually follow fishing boats in search of preys and does not prey on fish caught in artisan fishing gears. However, the extensive development of commercial fisheries in some parts of its distribution range have had an unknown effect on the amount of available food. Trawl fisheries are a source of entanglements and direct mortality, and incidental captures in shark nets have been reported for Uruguay (Scialabba 1989; Campagna 2008).

Conservation Genetics and Population Structure

Morphological and genetic studies support the idea that Peruvian colonies should be considered an ESU isolated from Atlantic populations: Túnez et al. (2007) using mtDNA cytb sequences analyzed the genetic diversity of the species in Cabo Polonio, Uruguay, and compared their results with available sequences from the Peruvian colony of Punta San Juan (Wynen et al. 2001).

No haplotypes were shared between colonies. Oliveira et al. (2008) combined morphometric and genetic data to compare the same Peruvian colony with individuals belonging to the Uruguayan breeding area. Seven polymorphic microsatellite loci reveal highly significant differences in allele frequencies between Uruguay and Peru. Morphological traits also supported these genetic differences. Altogether, these results suggest the existence of a different ESU in each ocean basin. More recently, Túnez et al. (2013) analyzed sequences of the mtDNA control region in the two southernmost breeding areas of the species, Uruguay and southern Chile-Staten Island. No haplotypes were shared between breeding areas, suggesting a long period of isolation between them and the possible existence of two different CUs.

Oliveira et al. (2006), using the same microsatellite loci used in their first study, found that the Peruvian population may have experienced a genetic bottleneck caused by the intense harvesting and the severe population reduction during the 1997-1998 El Niño event.

Southern Elephant Seal, *Mirounga Leonina*

Geographic Range, Population Numbers and Trends

Southern elephant seals have a circumpolar distribution in the Southern Hemisphere, occupying the island portion of Antarctic and Sub Antarctic waters (Perrin et al. 2009). The species breeds mainly on sub Antarctic islands, such as South Georgia, were approximately 54% of the world population reproduce (Boyd et al. 1996). Total population of the species was estimated in 664,000 individuals, distributed in 14 colonies (Laws 1994). Only two of them are located in southern South America, in Malvinas (Falkland) Islands and Peninsula Valdés, Chubut, Argentina. Malvinas population has been estimated in 1827 individuals that are 1 year old or older (Galimberti & Sanvito 2000). Peninsula Valdés population is much bigger and the only continental colony of the species. Its population was estimated in 42,000 individuals older than 1 year in 1997 (Lewis et al. 1998). Colonies in the South Atlantic ocean are stable (Laws 1994; Boyd et al. 1996), while those in the Southern Indian and Pacific Oceans have decreased by up to 50% (Campagna 2008). Peninsula Valdés population has grown between 1982 and 1992, with a 23% of increase in pup numbers (Campagna & Lewis 1992) and now is stable (Lewis et al. 1998).

Conservation Status

The species is currently categorized as “least concerned” by the IUCN and listed in CITES Appendix II. Any future exploitation within the Antarctic Treaty area (south of 60ºS) would be regulated by the Convention for the Conservation of Antarctic Seals. The Malvinas Islands Dependencies Conservation Ordinance provides protection for southern elephant seals on South Georgia and the South Sandwich Islands (Reijnders et al. 1993).

Habitat and Ecology

The species shows an extreme sexual dimorphism in size, possibly the largest of any mammal, with males five to six times heavier than the females (Perrin et al. 2009). Adult males typically reach 4.5 m in length and weigh 1,500-3,000 kg, while females are about 2.60-3.00 m long and weigh 350-600 kg. Newborn pups are about 1.3 m and 40-50 kg (Campagna 2008). Female elephant seals attain its sexual maturity at 3-5 years, while males reach it at 4 years (Perrin et al. 2009). Ninety percent of males die before the age of 10 years while 90% of females die before the age of 14 years (McCann 1980; Reijnders et al. 1993).

Southern elephant seals spend a large percentage of their lives at sea and only return to land to give birth, breed and moult. They prefer sandy and cobble beaches, but will haul-out on sea ice, snow and rocky terraces. Breeding season starts in August and last to late-October (Campagna 2008). Males arrive to the colonies earlier than females and fight for control of territories (Jones 1981). Pregnant females arrive later, and usually give birth within five days of their arrival to the colony, most of them at the beginning of the breeding season (McCann 1980). Females come into estrous about 4 days before they wean their pups (Campagna 2008). At Peninsula Valdés, females copulate a mean of 2.5 times within the 20 days after parturition, with a peak during the third week of October (Campagna et al. 1993).

The species is extremely polygynous. Males display agonistic interactions, establishing dominance hierarchies (McCann 1981). Dominant bulls monopolize access to most of females in the breeding colony, with harem sizes of about 120 to 150 females.

At sea, they range far from their colonies and predominantly feed between the Sub Antarctic convergence and the northern edge of the pack ice, south of the Antarctic convergence. Adult males typically venture further south than females (Campagna 2008). Dive depth and duration vary during the year and between sexes. Satellite tracking revealed that seals dive at depths between 400 and 1,000 m from more than 20 minutes each time (McIntyre et al. 2010).

Main preys taken are squid and fish (Daneri & Carlini 2002). Antarctic Notothenia fishes are thought to be important prey when these seals are near the Antarctic continental shelf (Daneri & Carlini 2002).

Major Threats

The species has been harvested for several thousand years by aboriginal and natives in Tasmania and New Zealand (Smith 1989; Bryden et al. 1999). After their near extinction due to commercial hunting in the 19th century, worldwide population of southern elephant seals have recovered dramatically in number, but some populations are now declining (Perrin et al. 2009).

The species has minimal interactions with commercial fisheries, although intensive fishing could potentially deplete important prey stocks. There is no evidence that recent declines for animals breeding in the Indian and Pacific Oceans are related to fisheries in the Southern Ocean (SCAR 1991). However, development of new fisheries at high latitudes could have a significant impact on elephant seal populations (SCAR 1991, Reijnders et al. 1993).

Conservation Genetics and Population Structure

Since 1993, several studies have analyzed the population genetic structure of the species. Slade et al. (1998) analyzed mitochondrial and nuclear DNA variation in three main populations in the south Atlantic, south Indian, and south Pacific oceans, and the smaller continental population in Peninsula Valdés. Their results indicate that population structure of mtDNA was strong. Nuclear DNA loci, although less informative, were consistent with the mtDNA results. Hoelzel et al. (2001) also compared mitochondrial and nuclear markers and morphometric measures between the Argentinean mainland population and South Georgia Island. They found pronounced differentiation at the mtDNA control region. Some morphometric characters and seven of the ten nuclear DNA markers also showed differentiation between the island and mainland sites. The same results were found by Fabiani et al. (2003) for mitochondrial markers, while results at nuclear ones were not coincident. They found strong evidence for male-mediated genetic dispersal over a geographic range of about 8000 km. In summary, southern elephant seals show current gene flow between distant populations, but fixed differences and reciprocal monophyly for mitochondrial markers, along with significant differences for some nuclear markers, which suggests the existence of different ESUs along the species distribution.

Southern Right Whale, *Eubalaena australis*

Geographic Range, Population Numbers and Trends

Southern right whales are found in all the three ocean basins in the Southern Hemisphere between 20º and 60ºS in latitude. During breeding season (austral winter) they congregate along nearshore areas off southern Australia, New Zealand, Atlantic coast of South America (Argentina and Brazil), and southern Africa. Small numbers are also seen off central Chile, Peru, Tristan da Cunha and the east coast of Madagascar (Reilly et al. 2013). In austral summer, southern right whales are found mainly in latitudes 40-50°S (Ohsumi & Kasamatsu 1986) but have been seen in the Antarctic as far south as 65°S (Bannister et al. 1999) and around South Georgia (Rowntree et al. 2001). Migration routes between grounds may include north-south as well as east-west movements.

It was extensively hunted during the whaling years and its populations depleted. Since the cessation of whaling, population sizes are gradually increasing, especially in Argentina off Península Valdés, Chubut Province (Payne et al. 1990), Uruguay (Costa et al. 2007) and southern Brazil, where the increase has been 14% per year from 1987 to 2003 (Groch et al. 2005). There has also been an increase in other Argentine provinces: Río Negro (Failla et al. 2008) and Santa Cruz (Belgrano et al. 2008).

Conservation Status

Although the species is still scarce relative to its historic abundance, is not considered under threat at the hemispheric level. It is currently categorized as "least concerned" by the IUCN and is included in CITES Appendix I and the Convention of Migratory Species (CMS). Some breeding populations, in particular that off Chile/Peru, are still very small and data are insufficient to determine whether they are recovering.

Habitat and Ecology

It is a large, stocky, baleen whale that weights between 80 to 100 tonnes, and average 14-15 m in length with females slightly larger than males. They are identified by their predominantly black colouration, lack of dorsal fin, large, broad flippers and 'callosity' patterns on their heads (raised patches of roughened skin covered by dense colonies of 'whale lice'/crustaceans) (Payne et al. 1983). Their blow is wide, V-shaped and up to 5m high. Commonly observed behaviours include breaching, waving and slapping their flippers on the sea surface, sailing (using the flukes to sail in the wind), lobtailing and spyhoping.

Southern right whales live in groups of up to 12 individuals, but are more commonly found in groups of two or three, unless at feeding grounds. Their distribution in summer is likely linked to the distribution of their principal prey species, primarily large zooplankton such as euphausiids (krill) and large-size copepods (Leaper et al. 2006, Tormosov et al. 1998). They show maternally inherited site fidelity to near-shore winter nursery grounds, but exactly where they feed in summer remains a mystery. Females usually calve once every 3 years (Cooke et al. 2003) and are not often seen on the calving grounds in intervening years (Payne 1986). This period includes one year in gestation (Best 1994), one year in lactation and one year recovering energy reserves to support the next pregnancy (Burnell 2001).

Major Threats

Southern right whales were intensively hunted throughout the Southern Hemisphere to near commercial extinction from the 18th through the early 20th centuries by American and European commercial whalers. Before whaling, total population was estimated in 55,000-70,000 whales (IWC 2001). Only between 1805 and 1844, about 45,000 right whales were killed, and the species came very close to extinction. Although internationally protected since 1935, Tormosov et al. (1998) describe unreported Soviet catches of right whales in their pelagic summer grounds in the mid 20th century and illegal hunts along the southern Brazilian coast until 1973.

Like northern right whales, they are subject to mortality due to entanglements in fishing gear and collisions with shipping (IWC 2001). In the Península Valdés, Argentina calving ground, kelp gulls (*Larus dominicanus*) feed on pieces of skin and blubber they rip from the whales' backs, producing injuries and eventually driving whales elsewhere (Rowntree et al. 1998, Fazio et al. 2012). In the last ten years, hundreds of southern right whales calves have died on this calving ground, with no particular cause found (Rowntree et al. 2013).

Other threats include chemical contaminants such as metals due to an increase in population, tourism and industrial growth in the area (Rosas et al., 2012; Martino et al. 2013), changes in global climate signals (Leaper et al., 2006) and tourism activity (Lundquist et al. 2013). Observed correlations between breeding success off Argentina and sea surface temperature anomalies at South Georgia suggest that as Antarctic feeding grounds warm up, the average calving rate of southern right whales can be expected to decline. These results extend similar observations from other krill predators and show clear linkages between global climate signals and the biological processes affecting whale population dynamics (Leaper et al. 2006). According to Lundquist et al. (2013), increased levels of tourism activity are a concern, as reduction in resting time and disruption of socialization among adults, juveniles, and mother/calf pairs have unknown long-term consequences.

There are protected marine areas with specific management measures aimed at protecting the right whales in their calving grounds. These include the Area de Proteção Ambiental de Baleia Franca (Right Whale Environmental Protection Area) off Catarina State in Brazil, and the Golfo San José Provincial Marine Park (Parque Marino Golfo San José) in Argentina.

Conservation Genetics and Population Structure

Genetic and stable isotope studies have provided insight into the species foraging ecology and population structure (Best & Schell 1996, Rowntree et al., 2008, Valenzuela et al. 2009, 2010). The overall genetic differentiation among southern right whale nursery areas has been well documented and indicates little exchange of females between calving grounds, which could be explained from maternal fidelity to nursery grounds (Portway 1998, Baker et al. 1999; Patenaude et al. 2007; Valenzuela et al. 2010). This was also shown with long-term photo identification studies (Best el al. 1993). Within ocean basins, no genetic differentiation has been found between nursery grounds and feeding grounds (Portway 1998; Baker et al. 1999; Patenaude et al. 2007). As pointed out by Portway (1998), commercial whaling appears not to have reduced genetic diversity in the Península Valdés population to an extent that has affected its recovery. Valenzuela et al. (2009) reported that genetic heterogeneity on the nursery ground at Península Valdés appears to be driven by population structure on the feeding grounds. This is the first indication of genetic heterogeneity within a nursery-area subpopulation of southern right whales. The consequences of such substructuring within a breeding population are unknown; however, a likely result is that the effects of changes in food abundance within a particular feeding ground might not be spread across an entire breeding population, but instead focused on particular genetic lineages (Valenzuela et al. 2009).

Commerson's Dolphin, *Cephalorhynchus Commersonii commersonii*

Geographic Range, Population Numbers and Trends

The subspecies is found in shallow waters of the continental shelf along the southeastern coast of South America, with northern limit near Rio Negro mouth (40ºS) and southern limit near Cape Horn (56ºS), including the central and eastern Strait of Magellan and the Malvinas (Falkland) Islands (Goodall et al., 1988, Goodall 1994). In the southwestern coast of South America, it is found in the Strait of Magellan, Seno Skyring, Fitz Roy channel, seno Otway and seno Almirantazgo (Goodall et al. 1988, Aguayo-Lobo et al. 1998, Gibbons et al. 2000). Vagrants have been reported in Brazil, Argentina and Chile (Brownell & Praderi 198;, Iñiguez et al. 2010; Pinedo et al. 2002).

Commerson's dolphin seems to be the most abundant species of the genus *Cephalorhynchus* (Dawson 2002) although much of its range has not been surveyed and there are only a few estimates of abundance. Recent aerial surveys suggest that there are approximately 21,000 animals along the entire South American coast, with 7,000 between 42-48ºS and 14,000 in Tierra del Fuego (Leatherwood et al. 1988; Pedraza 2008).

Conservation Status

The species is currently categorized as "data deficient" by the IUCN and included in CITES Appendix II, as a species not necessarily threatened but that may become so if not closely monitored. The South American subspecies is also listed in Appendix II of CMS.

Habitat and Ecology

The Commerson's dolphin is one of the smallest cetaceans in the world, with a maximum total length of 1.48 m and a weight of up to 45 kg. It has a distinctive pigmentation pattern in adults, with a black head, dorsal fin and fluke, and a white throat and body. Calves have a grey colour in their body, later turning into white in the adult form. The head is blunt (with no beak) and has a stocky body with rounded flippers. It is possible to identify sexes with the genital patch, which is shaped like a raindrop in males, and more rounded in females with two lobules around the mammary slits.

Breeding season is in the austral spring and summer, from September to February (Jefferson et al. 1993; Goodall 1994). Gestation lasts between 10 and 11 months. Females bear their first calf at around 6-9 years, and males reach sexual maturity at 5-9 years (Lockyer et al. 1981; Lockyer et al. 1988; Goodall

1994). The maximum age found to date for the species is 24 years (Dellabianca 2011).

The Commerson's dolphin is a coastal dolphin, which inhabits the shallow waters of the continental shelf with a foraging habitat near shore shelf. They are found in inshore waters, particularly around kelp beds in sheltered bays and harbours (Goodall et al. 1988, Leatherwood et al. 1988, Lescrauwaet et al. 2000). It appears to move seasonally in small groups (2–10 individuals) between inshore and offshore habitats and make diurnal movements closely related to the distribution of their prey (Goodall 1994; Pedraza 2008; Coscarella et al. 2010; Garaffo et al. 2011). The diet studies indicated that it is a generalist feeder including alternative prey such as fishes, squids, and invertebrates (Bastida et al. 1988; Riccialdelli et al. 2010). Isotopic comparisons suggest that all age classes share the same foraging areas (Riccialdelli et al. 2013).

Major Threats

The earliest records show the presence of Commerson's dolphin bones in kitchen middens of Tierra del Fuego old natives, Yamana or Yagan (Piana et al. 1985). Across some of its range, it have historically been subject to directed exploitations for use as bait in the crab (*Lithodes antarctica*) fishery, as well as for oil, sport, live capture for exhibition, scientific specimens and even for human consumption (Goodall & Cameron 1980; Goodall et al. 1988, 1994). Although this directed exploitation no longer occurs, the species is still being caught incidentally in trammel nets, mid-water trawls and artisanal shore-based gillnet fisheries, especially along the shallow coastline of northern Tierra del Fuego and southern Patagonia, Argentina (Goodall et al. 1994; Crespo et al. 1997; Crespo et al. 2000; Iñiguez et al. 2003). Other threats to this species could include the decrease of key prey species due to uncontrolled fishing and factories, pollution, infectious diseases, habitat degradation in coastal waters, climate oscillations, oil drilling and production, and commercial shipping traffic throughout its range (Coscarella et al. 2003; Skewgar et al. 2007; Dellabianca et al. 2012; Cáceres-Saez et al. 2013; Roe et al. 2013).

Conservation Genetics and Population Structure

Recent genetic studies have evaluated the population structure of Commerson's dolphin and revealed significant differentiation among different studied areas (within Tierra del Fuego and Santa Cruz provinces) over small geographic scales, considering these as subpopulations or different management units (Pimper et al. 2010; Cipriano et al. 2011). Two "ecological stocks" have been identified in central Patagonia and Tierra del Fuego based on differences in parasite loads and patterns of prey consumption (Berón-Vera et al. 2001).

Given their limited dispersal in South America (Hevia et al., 2005, Coscarella et al. 2007; Pimper et al. 2010) and restricted near-shore distribution, it is likely that the Malvinas (Falkland) population represents a genetically isolated stock.

Although the analysis of mtDNA variation and distribution was highly informative for examining gene flow over these areas, other lines of evidence (microsatellite loci, SNPs or morphometric/morphological traits) are needed to improve knowledge of ESUs for Commerson's dolphins all along their distribution.

Franciscana, *Pontoporia blainvillei*

Geographic Range, Population Numbers and Trends

The franciscana, also known as the La Plata River dolphin, is a small cetacean endemic to the western South Atlantic Ocean that inhabits a narrow strip of coastal waters off Argentina, Uruguay and Brazil, extending from Itaúnas (18° 25′S) to Golfo Nuevo (42° 35′S) (Crespo et al. 1998; Siciliano et al. 2002). Four management areas (Franciscana Management Area or FMA) have been recognized in the entire distribution range of franciscana (Secchi et al. 2003): FMA I - coastal waters of Espírito Santo and Rio de Janeiro states, Brazil; FMA II - São Paulo, Paraná and Santa Catarina states, Brazil; FMA III - coastal waters of Rio Grande do Sul State, southern Brazil and Uruguay; and FMA IV - coastal waters of Argentina, including the provinces of Buenos Aires, Rio Negro and Chubut.

Until 2004, only two surveys were carried out to estimate the species density and abundance in the FMAs, but due to the methods employed and limitations of them, these were not consider suitable for providing a minimum estimate of abundance (IWC 2005). The most recent population estimate for this species in Argentinean waters (FMA IV) is 14,000 individuals (Crespo et al. 2010).

Conservation Status

The species is currently categorized as "vulnerable" by the IUCN and included in CITES Appendix II.Habitat and Ecology

It is one of the smallest dolphins, with females larger than males, ranging for adult specimens between 1.20 and 1.77 m and a weight between 29 and 53 kg (Pinedo et al. 1989). The species is brownish to dark grey and has an extremely long and narrow beak and a bulky head.

These dolphins are difficult to observe in the wild and tend to be found in small groups, from two or three individuals (Bordino et al. 1999), with many authors supporting the hypothesis of a single male mating system, a rare occurrence among mammals (Danilewicz et al. 2004, Panebianco et al. 2012, Wells et al. 2013). Age at sexual maturity is estimated to be around 2.7 years, and the gestation period is between 10.5 and 11.1 months. Breeding season is between November and March. The calving interval is around 2 years. Longevity has been estimated to be close to 15 years for males and 21 for females (Bordino et al. 1999; Rodriguez et al. 2002; Denuncio et al. 2013).

The franciscana feeds mostly near the bottom on the most abundant species in the region and seems to change its diet according to seasonal prey fluctuations. The diet also includes squids, octopus, and shrimps (Rodriguez et al. 2002). It is generally found in marine waters and occasionally in estuaries, within the 30 nautical miles and mainly in turbid waters < 30m deep (Pinedo et al. 1989).

Major Threats

The franciscana is considered to be the most threatened cetacean in the southwest Atlantic (Secchi 2010). Its coastal distribution greatly overlaps with fishing activities and bycatch in gillnets is the main threat to its conservation throughout their limited range (Praderi et al. 1989, Corcuera et al. 1994, Secchi 2010, among others). In Argentina, it is one of the most vulnerable cetaceans regularly entangled in coastal artisanal fishery nets (Crespo et al. 1994) resulting in the deaths of 400-600 individuals each year, at levels that are likely unsustainable (Bordino & Albareda 2004; Cappozzo et al. 2007; Negri et al. 2012). Based on a fisheries' monitoring program, Franco-Trecu et al. (2009) have estimated annual bycatch mortality in Uruguay of 289 franciscana dolphins. On the Brazilian coast, incidental mortality has been estimated in each local area from monitoring fisheries' programs and beach surveys, ranging from 25 bycaught individuals in Parana (Rosas et al. 2002) to 946 in Rio Grande do Sul (Secchi et al. 2004). Current estimates total at least

2,900 animals per year in all four management stocks, combined (Secchi et al. 2003), but methodologies applied in each area are different and the biases resulting from the lack of standardized criteria have not been evaluated.

Measures are needed to reduce the level of bycatch of this species, and research is needed to monitor bycatch levels more accurately (Reeves et al. 2012). The use of acoustic alarms has been shown to significantly reduce the bycatch rates of the species but financial costs and enforcement of pinger use limits their applicability in many small-scale fisheries (Read 2008). The use of modified gillnets (such as reflective or stiff nets) does not lead to a reduction in franciscana bycatch rates (Bordino et al. 2013). Unless alternative fishing grounds are identified and alternative gear adopted, populations of franciscana dolphin will inevitably lead to a decline in the near future (Negri et al. 2012).

Other potential threats include overexploitation and stock depletion of the main franciscana prey (Rodríguez et al. 2002), habitat degradation, heavy metals and plastic ingestion (Denuncio et al. 2011).

Conservation Genetics and Population Structure

Many studies have included analysis of mtDNA and microsatellites markers to better understand the population structure and define management stocks of the franciscana dolphin along its distribution. Secchi et al. (1998) found that at least two populations of the species exist north and south of Santa Catarina, Brazil. This was in accordance with two types of franciscana that Pinedo (1991) distinguished based on osteological differences. Lazaro et al. (2004) agreed with these findings and further suggested the presence of two additional populations, one in what they called Rio de la Plata (Uruguay and northern Buenos Aires combined) and the other in Claromeco, Argentina. Mendez et al. (2008) suggested that the population in northern Buenos Aires was the most isolated population in Argentina and argue for a finer-scale subdivision within this region (FMA IV), suggesting the existence of an estuary subpopulation differing from ocean populations. Later, combined evidence of genetic and environmental structure suggests that isolation by distance can explain the occurrence of genetically isolated populations within Argentina, in areas that are environmentally distinct (Mendez et al. 2010). Studies on both broad and fine scales have shown strong evidence of genetic structuring in franciscana dolphin populations.

Genetic data also confirm the association of mothers and their calves killed in the same fishing nets, killing not just individuals but functional social units (Valsecchi & Zanelatto 2003; Mendez et al. 2010; Costa-Urrutia et al. 2012; Wells et al. 2013). These authors have also suggested that captures of unrelated adult females and adult males in the same nets indicate the occurrence of breeding pairs leading to the suggestion of a single male or serial monogamy breeding system for this species.

REFERENCES

Aguayo-Lobo, A., Torres, D. & Acevedo, J. (1998). Los mamíferos marinos de Chile: I. *Cetacea, Serie Científica INACH 48*, 19-159.

Arias-Schreiber M., Rivas, C. (1998). Distribuición, tamaño y estructura de las poblaciones de lobos marinos *Arctocephalus australis* y *Otaria byronia* en el litoral Peruano, en Noviembre 1996 y Marzo 1997. *Informe de los Progresos del Instituto del Mar del Perú 73*, 17-32.

Baker C. S., Patenaude N. J., Bannister, J. L., Robins J. & Kato, H. (1999). Distribution and diversity of mtDNA lineages among southern right whales (*Eubalaena australis*) from Australia and New Zealand. *Marine Biology 134*, 1-7.

Bannister J. L., Pastene L. A. & Burnell, S. R. (1999). First record of movements of a southern right whale (*Eubalaena australis*) between warm water breeding grounds and the Antarctic Ocean, South of 60ºS. *Marine Mammal Science 15*, 1337-1342.

Bastida, R., Lichtschein, V. & Goodall, R. N. P. (1988). Food habits of *Cephalorhynchus commersonii* off Tierra del Fuego. In: *International Whaling Commission, Special Issue, v. 9*, pp 3-70.

Batallés, M., Pin, O. & Lima, M. (1990). Estudio del crecimiento del lobo fino sudamericano (*Arctocephalus australis*) en Isla de Lobos, Uruguay. *Frente Marítimo 7*, 69-73.

Belgrano J., Iñíguez, M. Gibbons, J., Garcia, C. & Olavarria, C. (2008). Southwest Atlantic right whales *Eubalaena australis* (Desmoulins, 1822) distribution nearby the Magellan Strait. *Anales Instituto Patagonia 36*, 69-74.

Beron-Vera, B., Pedraza, S. N., Raga, J. A., De Pertierra, A. G., Crespo, E. A., Alonso, M. A. & Goodall, R. N. P. (2001). Gastrointestinal helminths of Commerson's dolphins *Cephalorhynchus commersonii* from central Patagonia and Tierra del Fuego. *Diseases of Aquatic Organisms 47*, 201-208.

Best, P. B. (1994). Seasonality of reproduction and the length of gestation in southern right whales *Eubalaena australis*. *Journal of Zoology 232*, 175-189.

Best, P. B. & Schell, D. M. (1996). Stable isotopes in southern right whale (*Eubalaena australis*) baleen as indicators of seasonal movements, feeding and growth. *Marine Biology 124*, 483-494.

Best, P. B., Payne, R., Rowntree, V., Palazo, J. T. & do Carmo Both, M. (1993). Long-range movements of South Atlantic right whales *Eubalaena australis*. *Marine Mammal Science 9*, 227-234.

Bordino, P. & Albareda, D. (2004). La mortalidad incidental de delfines franciscana *Pontoporia blainvillei* en la pesca con redes de enmalle costeras en el norte de Buenos Aires, Argentina. Paper SC/56/SM11 presented to the IWC Scientific Committee.

Bordino, P., Thompson, G. & Iñiguez, M. (1999). Ecology and behaviour of the franciscana dolphin *Pontoporia blainvillei* in Bahía Anegada, Argentina. *Journal of Cetacean Research and Management*, 1: 213-222.

Bordino, P., Mackay, A. I., Werner, T. B., Northridge, S. P. & Read, A. J. (2013). Franciscana bycatch is not reduced by acoustically reflective or physically stiffened gillnets. *Endangered Species Research*, 21: 1-12.

Boyd, I. L., Walker, T. R. & Poncet, J. (1996). Status of southern elephant seals at South Georgia. *Antarctic Science*, 8: 237-244.

Brownell, R. L. Jr. & Praderi, R. (1985.) Taxonomy and distribution of Commerson's dolphin, *Cephalorhynchus commersonii*. *Scientific Report of the Whales Research Institute*, 36: 153-164.

Bryden, M. M., O'Connor, S. & Jones, R. (1999). Archeological evidence for the extinction of a breeding population of elephant seals in Tasmania in prehistoric times. *International Journal of Osteoarcheology 9*, 430-437.

Burnell, S. (2001). Aspects of the reproductive biology and behavioural ecology of right whales off Australia. *Journal of Cetacean Research and Management, (Special Issue) 2*, 89-102.

Cabrera A. & Yepes, J. (1940). *Mamíferos sudamericanos* Buenos Aires, Argentina: Ediciones Ediar.

Cáceres-Saez, I., Dellabianca, N. A., Goodall, R. N. P., Cappozzo, H. L. & Ribeiro Guevara, S. (2013). Mercury and Selenium in subantarctic commerson's dolphins (*Cephalorhynchus c. commersonii*). *Biological Trace Element Research 151*, 195-208.

Campagna C. (1985). The breeding cycle of the southern sea lion, *Otaria byronia*. *Marine Mammal Science 1*, 210-218.

Campagna C. (2008). IUCN SSC Pinniped Specialist Group. *IUCN Red List of Threatened Species*.

Campagna C. & Le Boeuf B. J. (1988). Reproductive behaviour of Southern sea lions. *Behaviour 104*, 233-261.

Campagna C. & Lewis, M. (1992). Growth and distribution of a southern elephant seal colony. *Marine Mammal Science 8*, 387-396.

Campagna C., Lewis M. & Baldi R. R. (1993) Breeding biology of southern elephant seals in Patagonia. *Marine Mammal Science 9*, 34-47.

Campagna C., Werner, R., Karesh, W., Marín, M. R., Koontz, F., Cook, R. & Koontz, C. (2001). Movements and location at sea of South American sea lions (*Otaria flavescens*). *Journal of Zoology 257*, 205-220.

Cappozzo, H. L. (1995). *Comportamiento reproductivo en dos especies de Otáridos de América del Sur.* PhD Thesis, Universidad de Buenos Aires, Buenos Aires.

Cappozzo, H. L. (2002). South American sea lion (*Otaria flavescens*). In Perrin, W. F., B. Wursig, & J. G. M. Thewissen (Eds.). *Encyclopedia of marine mammals* (pp 1143-1146). San Diego, USA: Academic Press.

Cappozzo, H. L., Campagna, C. & Monserrat, J. (1991). Sexual dimorphism in newborn southern sea lions. *Marine Mammal Science 7*, 385-394.

Cappozzo, H. L., Negri, M. F., Pérez, F. H., Alabareda, D., Monzón, F. & Corcuera, J. F. (2007). Incidental mortality of franciscana dolphin, (*Pontoporia blainvillei*) in Argentina. *Latin American Journal of Aquatic Mammals 6*, 127-137.

Cappozzo, H. L., Túnez, J. I. & Cassini, M. H. (2008). Sexual harassment and female gregariousness in the South American sea lion, *Otaria flavescens. Naturwissenschaften 95*, 625-630.

Cassini, M. H. (1999). The evolution of reproductive systems in pinnipeds. *Behavioral Ecology 10*, 612-616.

Cassini, M. H. (2000). A model on female breeding dispersion and the reproductive systems of pinnipeds. *Behavioural Processes 51*, 93-99.

Cipriano F., Hevia, M. & Iñíguez, M. (2011). Genetic divergence over small geographic scales and conservation implications for Commerson's dolphins (*Cephalorhynchus commersonii*) in southern Argentina. *Marine Mammal Science 27*, 701-718.

Committee on Taxonomy (2011). List of marine mammal species and subspecies. Available at Society for Marine Mammalogy, www. marinemammalscience. org, consulted on November 20 2013.

Cooke J., Rowntree, V. J. & Payne, R. (2003). Analysis of inter-annual variation in reproductive success of South Atlantic right whales (*Eubalaena australis*) from photo-identifications of calving females observed off Península Valdés, Argentina, during 1971–2000. Paper SC/55/O23 presented to the IWC Scientific Committee, Berlin, June 2003 (unpublished). Available at iwcoffice.org.

Corcuera J., Monzon, F., Crespo, E. A., Aguilar, A. & Raga, J. A. (1994). Interactions between marine mammals and the coastal fisheries of Necochea and Claromeco (Buenos Aires Province, Argentina). *Reports of the International Whaling Commission, Special Issue 15,* 283-290.

Coscarella, M. A., Dans, S. L., Crespo, E. A. & Pedraza, S. N. (2003). Potential impact of unregulated dolphin watching activities in Patagonia. *Journal of Cetacean Research and Management 5*, 77-84.

Coscarella, M. A., Hoelzel, R., Crespo, E. A. & Loizaga de Castro, R. (2007). Population genetic structure of Commerson's dolphin along the continental Patagonian coast based on mtDNA. *21st Conference of the European Cetacean Society. Abstracts.*

Coscarella, M. A., Pedraza S. N. & Crespo, E. A. (2010). Behaviour and seasonal variation in the relative abundance of Commerson's dolphin (*Cephalorhynchus commersonii*) in northern Patagonia, Argentina. *Journal of Ethology 28*, 463-470.

Costa P., Piedra, M., Franco, P. & Paez, E. (2007), Distribution and habitat use patterns of southern right whales, *Eubalaena australis*, off Uruguay. *Journal of Cetacean Research and Management*, 9: 45-51.

Costa-Urrutia P., Abud, C., Secchi, E. R. & Lessa, E. P. (2012). Population genetic structure and social kin associations of franciscana dolphin, *Pontoporia blainvillei. Journal of Heredity 103*, 92-102.

Crespo, E. A. (1988). *Dinámica poblacional del lobo marino del sur Otaria flavescens (Shaw 1800), en el norte del litoral patagónico*. PhD Thesis, Universidad de Buenos Aires, Buenos Aires.

Crespo, E. A. & Pedraza, S. N. (1991). Estado actual y tendencia de la población de lobos marinos de un pelo (*Otaria flavescens*) en el litoral norpatagónico. *Ecología Austral 1*, 87-95.

Crespo, E. A., Corcuera, J. F. & López Casorla, A. (1994). Interactions between marine mammals and fisheries in some coastal fishing areas of Argentina. *Reports of the International Whaling Commission, Special Issue 15*, 269-281.

Crespo, E. A., Pedraza, S.N., Dans, S. L., Koen-Alonso, M., Reyes, L. M., García, N. A., Coscarella, M. & Schiavini, A. C. M. (1997). Direct and indirect effects of the highseas fisheries on the marine mammal populations in the northern and central Patagonian coast. *Journal of Northwest Atlantic Fishery Science 22*, 189-207.

Crespo, E. A., Harris, G. & Gonzalez, R. (1998). Group size and distributional range of the franciscana, *Pontoporia blainvillei. Marine Mammal Science 14*, 845-848.

Crespo, E. A., Schiavini, A. C. M., Pérez, F. H. & Cappozzo, H. L. (1999). Distribution, abundance and seasonal changes of South American fur seals, *Arctocephalus australis*, along the coasts of Argentina. In: *Proceedings of the 13th Annual Conference of the European Cetacean Society*, 36-37.

Crespo, E. A., Koen Alonso, M., Dans, S. L., García, N. A., Pedraza, S. N., Coscarella, M. A. & González, R. (2000). Incidental catch of dolphins in mid-water trawls for southern anchovy off Patagonia. *Journal of Cetacean Research and Management* 2, 11-16.

Crespo, E. A., Pedraza, S. N., Grandi, M. F., Danz, S. L. & Garaffo, G. V. (2010). Abundance and distribution of endangered franciscana dolphins in argentine waters and conservation implications. *Marine Mammal Science 26*, 17-35.

Daneri, G. & Carlini, A. (2002). Fish prey of southern elephant seals, *Mirounga leonina*, at King George Island. *Polar Biology 25*, 739-743.

Danilewicz, D., Claver, J. A., Perez-Carrera, A. L., Secchi, E. R & Fontoura, N. F. (2004). Reproductive biology of male franciscanas (*Pontoporia blainvillei*) (Mammalia, Cetacea) from Rio Grande do Sul, Southern Brazil. *Fishery Bulletin 102*, 581-592.

Dans, S. L., Koen-Alonso, M., Crespo, E. A., Pedraza, S. N. & García, N. (2003). Interactions between marine mammals and high seas fisheries in Patagonia under an integrated approach. In: Gales N., M. Hindell & R. Kirkwood (Eds.). *Marine Mammals: Fisheries; Tourism And Management Issues* (pp 100-115). Collingwood, Australia: CSIRO Publishing.

Dans, S. L, Crespo, E. A., Pedraza, S. N. & Koen-Alonso, M. (2004). Recovery of the South American sea lion (*Otaria flavescens*) population in northern Patagonia. *Canadian Journal of Fisheries and Aquatic Science 61*, 1681-1690.

Dawson, S. M. (2002). Cephalorhynchus dolphins *Cephalorhynchus* spp. In: Perrin, W. F., B. Wursig, & , J. G. M. Thewissen (Eds). *Encyclopedia of Marine Mammals* (pp 200-204). San Diego, USA: Academic Press.

Denuncio, P., Bastida, R., Dassis, M., Giardino, G., Gerpe, M. & Rodríguez, D. (2011). Plastic ingestion in Franciscana dolphins, *Pontoporia blainvillei* (Gervais and d'Orbigny, 1844), from Argentina. *Marine Pollution Bulletin 62*, 1836-1841.

Denuncio, P., Bastida, R., Danilewicz, D., Morón, S., Rodríguez-Heredia, S. & Rodríguez, D. H. (2013). Calf chronology of the franciscana dolphin (*Pontoporia blainvillei*): birth, onset of feeding, and duration of lactation in coastal waters of Argentina. *Aquatic Mammals 39*, 21-28.

Dellabianca, N. A. (2011). *Efectos de ciclos climáticos sobre la tonina overa,* Cephalorhynchus commersonii *en aguas subantárticas.* Ph.D. Thesis, Universidad de La Plata.

Dellabianca, N. A., Hohn, A. A., Goodall, R. N. P., Pousa, J. L., Macleod, C. D & Lima, M. (2012). Influence of climate oscillations on dentinal deposition in teeth of Commerson's dolphins. *Global Change Biology 18*, 2477-2486.

Fabiani A., Hoelzel A. R., Galimberti F. & Muelbert M. M. C. (2003) Long-range paternal gene flow in the southern elephant seal. *Science 299*, 676.

Failla, M., Vermuelen, E., Carabajal, M., Arruda, J., Godoy, H., Lapa, A., Mora, G., Urrutia, C., Balbiano, A. & Cammareri, A. (2008). Historical records of southern right whales (*Eubalaena australis*) of the province of Río Negro, North Patagonia, Argentina (1991-2008). Paper SC/60/BRG1 presented to the IWC Scientific Committee, June 2008 (unpublished), 9 pp (Available in iwcoffice.org.).

Fazio A., Bertellotti M., Villanueva. C. (2012) Kelp gulls attack Southern right whales: a conservation concern? *Marine Biology 159*, 1981-1990.

Feijoo M., Lessa E. P., Loizaga de Castro R. & Crespo E. A. (2011) Mitochondrial and microsatellite assessment of population structure of South American sea lion (*Otaria flavescens*) in the Southwestern Atlantic Ocean. *Marine Biology 158*, 1857-1867.

Franco-Trecu, V., Costa, P., Abud, C., Dimitriadis, C., Laporta, P., Passadore, C. & Szephegy, M. (2009). By-catch of franciscana *Pontoporia blainvillei* in uruguayan artisanal gillnet fisheries: an evaluation after a twelve-year gap in data collection. *Latin American Journal of Aquatic Mammals 7*, 11-22.

Galimberti, F. & Sanvito, S. (2000). *Research on elephant seals of Sea Lion Island: annual report 1999.* Unpublished report. Available online in http://web.tiscalinet.it/esrg/pdf_vari/rep_1999.pdf)

Garaffo, G. V., Dans S. L., Pedraza, S. N., Degrati, M., Schiavini, A., González, R. & Crespo E. A. (2011). Modeling habitat use for dusky dolphin and Commerson's dolphin in Patagonia. *Marine Ecology Progress Series 421*, 217-227.

Gibbons, J., Gazitúa, F. & Venegas, C. (2000), Cetáceos en el Estrecho de Magallanes y Senos Otway, Skyring y Almirantazgo. *Anales Instituto Patagonia, Serie Ciencias Naturales 28*, 107-118.

Goodall, R. N. P. (1994). Commerson's dolphin, *Cephalorhynchus commersonii* (Lacépede 1804). In: Ridgway S & R. J. Harrison (Eds.).

Handbook of marine mammals, The First Book of Dolphins, *vol 5* (pp. 241-67). London, UK: Academic Press.

Goodall, R. N. P. & Cameron, I. S. (1980). Exploitation of small cetaceans off southern South America. *Reports of the International Whaling Commission 30*, 445–450.

Goodall, R. N. P., Galeazzi, A. R., Leatherwood S., Miller, K. W., Cameron, I. S., Kastelein R. A. & Sobral, A. P. (1988). Studies of Commerson's dolphins, *Cephalorhynchus commersonii*, off Tierra del Fuego, 1976–1984, with a review of information on the species in the South Atlantic. In: Brownell, R. L. Jr & G. P. Donovan (Eds). *Biology of the genus* Cephalorhynchus, *Reports of the International Whaling Commission, Special Issue 9* (pp 3–70). Cambridge, UK.

Goodall, R. N. P., Schiavini, A. C. M., Fermani, C. (1994). Net fisheries and net mortality of small cetaceans off Tierra del Fuego, Argentina. In: W. F. Perrin, G. P. Donovan & J.Barlow (Eds.). *Gillnets and Cetaceans Report of the International Whaling Commission, Special Issue 15* (pp. 295-304).

Groch, K., Palazzo Jr., J. T., Flores, P. A. C., Adler, F. R. & Fabian, M. E. (2005). Recent rapid increases in the right whale (*Eubalaena australis*) population off southern Brazil. *Latin American Journal of Aquatic Mammals 4*, 41-47.

Hevia, M., Iniguez, M., Tossenberger, V. & Cipriano, F. (2005). Genetic divergence between Commerson's dolphin sub-populations in southern Argentina. *16th Biennial Conference on the Biology of Marine Mammals*. San Diego, USA: Society of Marine Mammalogy Conference Abstracts.

Hoelzel, A. R., Campagna, C. & Arnbom, T. (2001). Genetic and morphometric differentiation between island and mainland southern elephant seal populations. *Proceedings of the Royal Society of London. Series B. Biological Sciences 268*, 325-332.

International Union for Conservation of Nature (2013). IUCN Red List of Threatened Species. Version 2013.2. Available at www.iucnredlist.org.

International Whaling Commission (2001) Report of the workshop on the comprehensive assessment of right whales: a worldwide comparison. *Journal of Cetacean Research and Management* 2, 1-60.

International Whaling Commission (2005) Annex L. Report of the Sub-Committee on Small Cetaceans. *Journal of Cetacean Research and Management 7*, 307-326.

Iñíguez, M. A., Hevia, M., Gasparrou, C., Tomsin, A. L. & Secchi, E. R. (2003). Preliminary estimate of incidental mortality of Commerson's dolphins (*Cephalorhynchus commersonii*) in an artisanal setnet fishery in

La Angelina beach and Ría Gallegos, Santa Cruz, Argentina. *Latin American Journal of Aquatic Mammals 2*, 87-94.

Iñíguez, M., Hevia, M., Cipriano, F., Sarradell, J. & Doumecq Milieu, R. (2010). Stranding of a Commerson's dolphin, *Cephalorhynchus commersonii*, in Buenos Aires Province, Argentina. *Marine Biodiversity Records 3*.

Jefferson, T. A., Leatherwood, S. & Webber, M. A. (1993). *FAO Species identification guide. Marine mammals of the world.* Rome, Italy: UNEP/FAO.

Jones, E. (1981), Age in relation to breeding status of the male Southern Elephant Seal, *Mirounga leonina* (L.), at Macquarie Island. *Australian Wildlife Research 8*, 327-334.

King, J. E. (1983), *Seals of the world.* Santa Lucia, Australia: University of Queensland Press.

Koen-Alonso, M., Crespo, E. A., Pedraza, S. N., García, N. A. & Coscarella, M. A. (2000). Food habits of the South American sea lion, *Otaria flavescens*, off Patagonia, Argentina. *Fisheries Bulletin*, 98: 250-263.

Laws, R. M. (1994). History and present status of southern elephant seal populations. In Le Boeuf, B. J. & R. M. Laws (Eds.). *Elephant seals: population ecology, behaviour, and physiology*, (pp 49-65). Berkley, USA: University of California Press.

Lazaro, M., Lessa, E. P. & Hamilton, H. (2004). Geographic genetic structure in the franciscana dolphin (*Pontoporia blainvillei*). *Marine Mammal Science 20*, 201-214.

Leaper, R., Cooke, J., Trathan, P., Reid, K., Rowntree, V. & Payne, R. (2006). Global climate drives southern right whale (*Eubalaena australis*) population dynamics. *Biology Letters 2*, 289-292.

Leatherwood S., Kastelein R. A. & Hammond, P. S. (1988) Estimate of number of Commerson's dolphins in a portion of the northeastern Strait of Magellan, January-February 1984. *Reports of the International Whaling Commission 9*, 93-102.

Lescrauwaet, A. C., Gibbons, J., Guzman, L. & Schiavini, A. (2000). Abundance estimation of Commerson's dolphin in the eastern area of the Strait of Magellan-Chile. *Revista Chilena de Historia Natural 73*, 473-478.

Lewis, M., Campagna, C., Quintana, F. & Falabella, V. (1998). Estado actual y distribución de la población del elefante marino del sur en la Península Valdés, Argentina. *Mastozoología Neotropical 5*, 29-40.

Lockyer, C., Smellie, G. G., Goodall, R. N. P. & Cameron, I. S. (1981). Examination of teeth of Commerson's dolphin, *Cephalorhynchus commersonii*, for age determination. *Journal of Zoology 195*, 123-131.

Lockyer, C., Goodall, R. N. P. & Galeazzi, A. R. (1988). Age and body length characteristics of *Cephalorhynchus commersonii* from incidentally caught specimens off Tierra del Fuego. In Brownell, Jr R. L. & G. P. Donovan (Eds.). *Report of the International Whaling Commission, Special Issue v. 9* (pp. 103-118). Cambridge, UK: International Whaling Commission.

Lundquist, D., Sironi, M., Wursig, B., Rowntree, V., Martino, J. & Lundquist, L. (2013). Responses of southern right whales to simulated swim-with-whale tourism at Peninsula Valdes, Argentina. *Marine Mammals Science 29*, 24-45.

Martino, J., Wise, S. S., Perkins, C., Sironi, M. & Wise, J. P. Sr (2013). Metal levels in southern right whales (*Eubalaena australis*) from Península Valdés, Argentina. *Journal of Environmental and Analytical Toxicology 3*, 190.

McCann, T. S. (1980). Population structure and social organization of southern elephant seals, *Mirounga leonina* (L.). *Biological Journal of the Linnean Society 14*, 133-150.

McCann T. S. (1981). Aggression and sexual activity of male southern elephant seals, *Mirounga leonina*. *Journal of Zoology 195*, 295-310.

McIntyre, T., de Bruyn, P. J. N., Ansorge, I. J., Bester, M. N., Bornemann, H., Plötz, J. & Tosh, C. A. (2010). A lifetime at depth: vertical distribution of southern elephant seals in the water column. *Polar Biology 33*, 1037-1048.

Mendez, M., Rosenbaum, H. C. & Bordino, P. (2008). Conservation genetics of the franciscana dolphin in Northern Argentina: population structure, by-catch impacts, and management implications. *Conservation Genetics 9*, 419-435.

Mendez, M., Rosenbaum, H. C., Subramaniam, A., Yackulic, C. & Bordino, P. (2010). Isolation by environmental distance in mobile marine species: molecular ecology of franciscana dolphins at their southern range. *Molecular Ecology 19*, 2212-2228.

Müller, G., Wilson, R. P., Quintana, F. & Campagna, C. (2006) Three-dimensional dive behavior of a South American Sea Lion (*Otaria flavescens*). In: Trites A. W., S.K. Atkinson, D. P. DeMaster, L. W. Fritz, T. S. Gelatt, L. D. Rea & K. M. Wynne (Eds.). *Sea lions of the world* (pp. 255-268). Fairbanks, USA: Sea Grant College Program, University of Alaska.

Naya, D. E., Arim, M. & Vargas, R. (2002). Diet of South American fur seals (*Arctocephalus australis*) in Isla de Lobos, Uruguay. *Marine Mammal Science 18*, 734-745.

Negri, M. F., Denuncio, P., Panebianco, M. V. & Cappozzo, H. L. (2012). Bycatch of franciscana dolphins *Pontoporia blainvillei* and the dynamic of artisanal fisheries in the species southernmost area of distribution. *Brazilian Journal of Oceanography 60*, 151-160.

Ohsumi, S. & Kasamatsu, F. (1986). Recent off-shore distribution of the southern right whale in summer. *Reports of the International Whaling Commission Special Issue 10*, 177-186.

Oliveira, L. R., Arias-Schreiber M., Meyer, D. & Morgante, J. S. (2006). Effective population size in a bottlenecked fur seal population. *Biological Conservation 131*, 505-509.

Oliveira, L. R., Hoffman, J. I., Hingst-Zaher, E., Majluf, P., Muelbert, M. M. C., Morgante, J. S. & Amos W. (2008). Morphological and genetic evidence for two evolutionarily significant units (ESUs) in the South American fur seal, *Arctocephalus australis*. *Conservation Genetics 9*, 1451-1466.

Páez, E. (2005). ¿Yo…? Otaria. ¿Y usted…?. In *III Jornadas de Conservación y Uso Sustentable de la Fauna Marina* (p. 28). Montevideo, Uruguay.

Panebianco, M. V., Negri, M. F. & Cappozzo, H. L. (2012). Reproductive aspects of male franciscana dolphins (*Pontoporia blainvillei*) off Argentina. *Animal Reproduction Science 131*, 41-48.

Patenaude, N. J., Portway, V. A., Schaeff, C. M., Bannister, J. L., Best, P. B., Payne, R. S., Rowntree, V. J., Rivarola, M. & Baker, C. S. (2007). Mitochondrial DNA diversity and population structure among southern right whales (*Eubalaena australis*). *Journal of Heredity 98*, 147-157.

Payne, R. (1986). Long-term behavioural studies of the southern right whale (*Eubalaena australis*). *Reports of the International Whaling Commission, Special Issue 10*, 161-167.

Payne. R., Brazier. O., Dorsey. E. M., Perkins, J. S., Rowntree, V. J. & Titus, A. (1983). External features in southern right whales (*Eubalaena australis*) and their use in identifying individuals. In: Payne, R. (Ed.). *Communication and behaviour of whales,* AAAS Selected Symposia 76 (pp. 371-445). Colorado, USA: Westview Press.

Payne, R., Rowntree, V. J., Perkins, J. S., Cooke, J. G. & Lankester, K. (1990). Population size, trends and reproductive parameters of right whales (*Eubalaena australis*) off Peninsula Valdes, Argentina. In: Hammond, P. S., S. A. Mizroch & G. P. Donovan (Eds.). *Individual recognition of*

Cetaceans: use of photo-identification and other techniques to estimates population parameters. Reports of the International Whaling Commission Special Issue 12, 271-278.

Pedraza, S. N. (2008). Ecología poblacional de la tonina overa *Cephalorhynchus commersonii* (Lacépède, 1804) en el litoral Patagónico. Ph.D. Thesis, Universidad de Buenos Aires.

Perrin, W. F., Würsig, B. & Thewissen, J. G. M. (2009). *Encyclopedia of Marine Mammals,* 2nd Edition. San Diego, USA: Academic Press.

Piana, E., Orquera, L., Goodall, R. N. P., Galeazzi, A. R. & Sobral, A. P. (1985). Cetacean remains in Beagle Channel middens. In: *Sixth Biennial Conference on the Biology of Marine Mammals* (p. 64). Vancouver, Canada.

Pimper, L. E., Baker, C. S., Goodall, R. N. P., Olavarría, C. & Remis, M. I. (2010). Mitochondrial DNA variation and population structure of Commerson's dolphins (*Cephalorhynchus commersonii*) in their southernmost distribution. *Conservation Genetics 11*, 2157-2168.

Pinedo, M. C. (1991). Development and variation of the franciscana *Pontoporia blainvillei*. Ph.D. thesis, University of California, Santa Cruz.

Pinedo, M. C., Praderi, R. & Brownell, Jr. R. L. (1989). Review of the biology and status of the franciscana *Pontoporia blainvillei.* In: Perrin, W. F., R. L. Jr.Brownell, K. Zhou & J. Liu (Eds.). *Biology and Conservation of the River Dolphins, Occasional Papers of the IUCN Species Survival Commission (SSC) 3*, 46-51.

Pinedo, M. C., Barreto, A. S., Lammardo, M. P., Andrade, A. L. V. & Geracitano, L. (2002). Northernmost records of the spectacled porpoise, Layard's beaked whale, Commerson's dolphin, and Peale's dolphin in the southwestern Atlantic Ocean. *Aquatic Mammals 28*, 32-37.

Portway, V. (1998). Genetic population structure of South Atlantic right whales (*Eubalaena australis*). Masters thesis, American University, Washington, DC.

Praderi, R., Pinedo, M. C. & Crespo, E. (1989). Conservation and management of *Pontoporia blainvillei* in Uruguay, Brazil and Argentina. In: Perrin, W. F., R. L. Jr.Brownell, K. Zhou & J. Liu (Eds.). Biology and Conservation of the River Dolphins, Occasional Papers of the IUCN Species Survival Commission (SSC) 3, 52-56.

Read, A. J. (2008). The looming crisis: interactions between marine mammals and fisheries. *Journal of Mammalogy 89*, 541-548.

Redford, K. H. & Eisenberg, J. F. (1992). *Mammals of the Neotropics, The Southern Cone: Chile, Argentina, Uruguay, Paraguay vol 2*. Chicago, USA. The University of Chicago Press.

Reeves, R., Dalebout, M., Jefferson, TA, Karkzmarski, L., Laidre, K., O'Corry-Crowe, G., Rojas-Bracho, L., Secchi, E., Slooten, E., Smith, B. D., Wang, J. Y., Zerbini, AN & Zhou, K. (2012). *Pontoporia blainvillei.* In: IUCN 2013. *IUCN Red List of Threatened Species. Version 2013.2.* www.iucnredlist.org.

Reijnders, P., Brasseur, S., van der Toorn, J., van der Wolf, P., Boyd, I., Harwood, J., Lavigne, D., & Lowry, L. (1993). *Seals, fur seals, sea lions, and walruses: status survey and conservation action plan*. Gland, Switzerland: IUCN/SSC Seal Specialist Group.

Reilly, S. B., Bannister, J. L., Best, P. B., Brown, M., Brownell, Jr. R. L., Butterworth, D. S., Clapham, P. J., Cooke, J., Donovan, G. P., Urbán, J. & Zerbini, A. N. (2013). *Eubalaena australis*. In: *IUCN 2013. IUCN Red List of Threatened Species. Version 2013.2.* www.iucnredlist.org.

Reyes, L. M., Crespo, E. A. & Szapkievich, V. (1999). Distribution and population size of the southern sea lion (*Otaria flavescens*) in central and southern Chubut, Patagonia, Argentina. *Marine Mammal Science 15*, 478-493.

Riccialdelli, L., Newsome, S. D., Fogel, M. L. & Goodall, R. N. P. (2010). Isotopic assessment of prey and habitat preferences of a cetacean community in the southwestern South Atlantic Ocean. *Marine Ecology Progress Series 418*, 235-248.

Riccialdelli, L., Newsome, S. D., Dellabianca, N. A., Bastida, R., Fogel, M. L. & Goodall, R. N. P. (2013). Ontogenetic diet shift in Commerson's dolphin (*Cephalorhynchus c. commersonii*) off Tierra del Fuego. *Polar Biology 36*, 617-627.

Riedman, M. L., (1990). *The Pinnipeds. Seals, sea lions and walruses.* Los Angeles, USA: University of California Press.

Rodríguez, D. & Bastida, R. (1998). Four hundred years in the history of pinniped colonies around Mar del Plata, Argentina. *Aquatic Conservation: Marine and Freshwater Ecosystems 8*, 721-735.

Rodríguez, D., Rivero L., & Bastida, R. (2002). Feeding ecology of the franciscana (*Pontoporia blainvillei*) in marine and estuarine waters of Argentina. *Latin American Journal of Aquatic Mammals 1*, 77-94.

Roe, W. D., Howe, L., Baker E. J., Burrows, L. & Hunter, S. A. (2013). An atypical genotype of *Toxoplasma gondii* as a cause of mortality in

Hector's dolphins (*Cephalorhynchus hectori*). *Veterinary Parasitology 192*, 67-74.

Rosas, F. C. W., Monteiro-Filho, E. L. A. & Oliveira, M. R. de (2002). Incidental catches of franciscana (*Pontoporia blainvillei*) on the Southern coast of São Paulo State and the coast of Paraná State, Brazil. *Latin American Journal of Aquatic Mammals 1*, 161-167.

Rosas, C. L., Gil, M. N. & Uhart, M. M. (2012). Trace metal concentrations in Southern Right whale (*Eubalaena australis*) at Península Valdés, Argentina. *Marine Pollution Bulletin 64*, 1255-1260.

Rowntree, V. J., Mcguinness, P., Marshall, K., Payne, R., Sironi, M. & Seger, J. (1998). Increase harassment of right whales (*Eubalaena australis*) by kelp gulls (*Larus dominicanus*) at Peninsula Valdez, Argentina. *Marine Mammal Science 14*, 99-115.

Rowntree, V. J., Payne, R. & Schell D. (2001). Changing patterns of habitat use by southern right whales (*Eubalaena australis*) identified on the nursery ground at Península Valdés, Argentina. *Journal of Cetacean Research and Management, Special Issue 2*, 133-144.

Rowntree, V. J., Valenzuela, L. O., Franco Fraguas, P. & Seger, J. (2008). Foraging behaviour of southern right whales (*Eubalaena australis*) inferred from variation of carbon stable isotope ratios in their baleen. *Paper SC/60/BRG23 presented to the IWC Scientific Committee.*

Rowntree, V. J., Uhart, M. M., Sironi, M., Chirife, A., Di Martino, M., La Sala, L., Musmeci, L., Mohamed, N., Andrejuk, J., McAloose, D., Sala, J. E., Carribero, A., Rally, H., Franco, M., Adler, F. R., Brownell, R. L., Jr, Seger, J. & Rowles, T. (2013). Unexplained recurring high mortality of southern right whale *Eubalaena australis* calves at Península Valdés, Argentina. *Marine Ecology Progress Series 493*, 275-289.

SCAR (1991) Report of the workshop on southern elephant seals, Monterey, California, May 22-23, 1991.

Schiavini, A., Crespo E. A. & Szapkievich, V. (2004). Status of the population of South American sea lion (*Otaria flavescens* Shaw, 1800) in southern Argentina. *Mammalian Biology 69*, 108-118.

Scialabba, N. (1989). World review of marine mammal entanglement in fishing gear and plastic marine debris. *Food and Agriculture Organization Report*.

Secchi, E. R. (2010). Review on the threats and conservation status of Franciscana, *Pontoporia blanvillei* (Cetacea, Pontoporiidae). In Shostell, J. M. & M. Ruiz-Garcia (Eds.) *Biology, evolution and conservation of river dolphins within South America and Asia vol 1* (pp 323-339). Hauppauge, USA: Nova Science Publishers Inc.

Secchi, E. R., Wang, J. Y., Murray, B. W., Rocha-Campos, C. C. & White, B. N. (1998). Population differentiation in the franciscana (*Pontoporia blainvillei*) from two geographic locations in Brazil as determined from mitochondrial DNA control region sequences. *Canadian Journal of Zoology 76*, 1622-1627.

Secchi, E. R., Danilewicz, D. & Ott, P. H. (2003). Applying the phylogeographic concept to identify franciscana dolphin stocks: implications to meet management objectives. *Journal of Cetacean Research and Management 5*, 61-68.

Secchi, E. R., Kinas, P. G. & Muelbert, M. (2004). Incidental catches of franciscana in coastal gillnet fisheries in the Franciscana Management Area III: period 1999-2000. *Latin American Journal of Aquatic Mammals 3*, 61-68.

Sepúlveda, M., Pérez, M. J., Sielfeld, W., Oliva, D., Durán, L. R., Rodríguez, L., Araos, V. & Buscaglia, M. (2007). Operational interaction between South American sea lions *Otaria flavescens* and artisanal (small-scale) fishing in Chile: Results from interview surveys and on-board observations. *Fisheries Research 83*, 332-340.

Sepúlveda, M., Oliva, D., Urra, A., Pérez-Álvarez, M. J., Moraga, R., Schrader, D., Hinostroza, P., Melo, A., Díaz, H., & Sielfeld, W. (2011). Distribution and abundance of the South American sea lion *Otaria flavescens* (Carnivora: Otariidae) along the central coast off Chile. *Revista Chilena de Historia Natural 84*, 97-106.

Siciliano, S., Di Beneditto, A. P. M. & Ramos R. M. A. (2002). A toninha, *Pontoporia blainvillei* (Gervais & d'Orbigny, 1844) (Mammalia, Cetacea, Pontoporiidae), nos estados do Rio de Janiero e Esprito Santo, costa sudeste do Brasil: Caracterizacao dos habitats e fatores de isolamento das populacoes. *Boletim do Museu Nacional 476.*

Sielfeld, W. (1999). The knowledge and conservation status of *Otaria flavescens* (Shaw, 1800) and *Arctocephalus australis* (Zimmermann, 1783) at the Chilean coasts. *Estudios Oceanológicos 18*, 81-96.

Skewgar, E., Dee Boersma, P., Harris, G. & Caille, G. (2007). Anchovy fishery threat to Patagonian ecosystem. *Science 315 (5808),* 45.

Slade, R. W., Moritz, C., Hoelzel, A. R. & Burton, H. R. (1998). Molecular population genetics of the southern elephant seal *Mirounga leonina. Genetics 149*, 1945-1957.

Smith, I. W. G. (1989). Maori impact on the marine megafauna: Pre-European distributions of New Zealand sea mammals. In: Sutton, D. G. (Ed.). *Saying so doesn't make it so* (pp 76-108). Auckland, New Zealand: New Zealand Archeological Association.

Szapkievich, V. B., Cappozzo, H. L., Crespo, E. A., Bernabeu, R. O., Comas, C. & Mudry, M. D. (1999). Genetic relatedness in two Southern sea lion (*Otaria flavescens*) rookeries in the southwestern Atlantic. *Mammalian Biology 64*, 246-250.

Szteren, D., Naya, D. E. & Arim, M. (2004). Overlap between pinniped summer diet and artisanal fishery catches in Uruguay. *Latin American Journal of Aquatic Mammals 3*, 119-125.

Thompson, D., Duck, C. D., McConnell, B. J. & Garrett, J. (1998). Foraging behaviour and diet of lactating female southern sea lions (*Otaria flavescens*) in the Falkland Islands. *Journal of Zoology, London*, *246*, 135-146.

Thompson, D., Moss, S. E. W. & Lovell, P. (2003). Foraging behaviour of South American fur seals *Arctocephalus australis*: extracting fine scale foraging behaviour from satellite tracks. *Marine Ecology Progress Series 260*, 285-296.

Thompson, D., Strange, I., Riddy, M. & Duck, C. D. (2005). The size and status of the population of Southern sea lions *Otaria flavescens* in the Falkland Islands. *Biological Conservation 121*, 357-367.

Tormosov, D. D., Mikhaliev, Y. A., Best, P. B., Zemsky, V. A., Sekiguchi, K. & Brownell, R. L. (1998). Soviet catches of southern right whales *Eubalaena australis* 1951–1971. Biological data and conservation implications. *Biological Conservation 86*, 185–197.

Túnez, J. I., Centrón, D., Cappozzo, H. L. & Cassini, M. H. (2007). Geographic distribution and diversity of mitochondrial DNA haplotypes in South American sea lions (*Otaria flavescens*) and fur seals (Arctocephalus *australis*). *Mammalian Biology 72*, 193-203.

Túnez, J. I., Cappozzo, H. L. & Cassini, M. H. (2008a). Natural and anthropogenic factors associated with the distribution of South American sea lion along the Atlantic coast. *Hydrobiologia 598*, 191-202.

Túnez, J. I., Cappozzo, H. L. & Cassini, M. H. (2008b). Regional factors associated with the distribution of South American fur seals along the Atlantic coast of South America. *ICES Journal of Marine Science 65*, 1733-1738.

Túnez, J. I., Cappozzo, H. L., Nardelli, M. & Cassini, M. H. (2010). Population genetic structure and historical population dynamics of the

South American sea lion, *Otaria flavescens*, in north-central Patagonia. *Genetica 138*, 831-841.

Túnez, J. I., Cappozzo, H. L., Pavés, H., Albareda, D. A. & Cassini, M. H. (2013). The role of Pleistocene glaciations in shaping the genetic structure of South American fur seals (*Arctocephalus australis*). *New Zealand Journal of Marine and Freshwater Research 47*, 139-152.

Valenzuela, L. O., Sironi, M., Rowntree, V. J. & Seger, J. (2009). Isotopic and genetic evidence for culturally inherited site fidelity to feeding grounds in southern right whales (*Eubalaena australis*). *Molecular Ecology 18*, 782-791.

Valenzuela, L. O., Sironi, M., Rowntree, V. J., La Sala, L. & others (2010). Population genetic structure of living and dead southern right whales (*Eubalaena australis*) off Península Valdés, Argentina. Paper SC/62/BRG15 presented to the IWC Scientific Committee.

Valsecchi, E. & Zanelatto, R. C. (2003). Molecular analysis of the social and population structure of the franciscana *Pontoporia blainvillei*: conservation implications. *Journal of Cetacean Research and Management 5*, 69-75.

Vaz-Ferreira, R. (1982). *Arctocephalus australis* (Zimmermann), South American fur seal. In *Mammals in the seas. Small cetaceans, seals, sirenians and otters (FAO y UNEP, Eds.), FAO Fisheries Series 5, vol 4* (pp. 497-508). Rome, Italy.

Vaz-Ferreira. R. & Ponce de León, A. (1987). South American fur seal, *Arctocephalus australis*, in Uruguay. In Croxall, J. P. & R. L. Gentry (Eds.), *Status, biology and ecology of fur seals (NOAA Technical Report, NMFS 51* (pp. 29-32).

Wells, R. S., Bordino, P. & Douglas, D. C. (2013). Patterns of social association in the franciscana, *Pontoporia blainvillei*. *Marine Mammal Science 29*, E520-E528.

Wynen, L. P., Goldsworthy, S. D., Insley, S. J., Adams, M., Bickham, J. W., Francis, J., Gallo, J. P., Hoelzel, A. R., Majluf, P., White, R. W. G. & Slade, R. (2001). Phylogenetic relationships within the eared seals (Otariidae: Carnivora): implications for the historical biogeography of the family. *Molecular Phylogenetics and Evolution 21*, 270-284.

Ximénez, I. M. & Langguth, E. (2002.) *Isla de Lobos*. Montevideo, Uruguay: Editorial Graphis.

York, A., Lima, M., Ponce de León, A., Malek, A. & Páez, E. (1998). First description of diving behavior of female South American fur seals in

Uruguay. In *The World Marine Mammal Science Conference (WMMSC)* (p. 153). Monaco.

CONCLUSION: NEOTROPICAL AND CARIBBEAN AQUATIC MAMMALS

A. Sebastián Muñoz, Christopher Markus Götz and Elizabeth Ramos Roca

The relationships between humans and aquatic mammals in the Neotropics varied across time and space, as the chapters of this book clearly show. It is noticeable, though, that a zooarchaeological approach to this long term relationships is still conditioned by the heterogeneity of the investigations across the different neotropical regions. This heterogeneity partly derives from three aspects that have to do with different research agendas carried out in the northern, central and southern neotropical areas covered by the book; the former are, in this sense, defined by the ways human occupation took place in the corresponding areas. For instance, the comparison among the topics addressed in the different chapters indicates that research goals formulated for past agricultural societies have emphasized an approach where the excavation techniques focus on the settlements; few areas outside the proper archaeological site limits are excavated, thus killing sites with the remains of heavy aquatic mammals could have been missed. Excavations at archaeological sites commonly aim at architectural structures, and faunal remains are usually recovered while documenting ancient floors, surfaces or middens. In Mesoamerica, for example, excavations decidedly located outside a pre-Colombian settlement, for example at a beach or cave, are still a relatively rare phenomenon, possibly caused by the nearly overwhelming abundance of structural remains from the ancient Mesoamerican civilizations. On the other hand, research goals put forward for the study of past hunter-gatherer societies like those that inhabited Patagonia have been usually build upon zooarchaeological evidence for the last 30 years, particularly to discuss human-animal interactions. As some of the papers of this volume show (see, the chapters of Borella and Zangrando) research on hunter-gatherers´ zooarchaeological record requires a methodological basis that has been growing at different levels, involving sampling techniques, research design and analytical tools. It is noticiable that the specifics of archaeological evidence of aquatic mammal are on the top of the list of that research agendas in the last decade. Yet, in spite of the indisputable scarceness of finds in some areas and differences in the research approaches and cultural systems at the areas treated in the book, there is clear archaeological evidence of the use of aquatic mammals by the diverse past human societies that inhabited the continent. One of the expected goals of the book is to incentivize investigators of the human-animal relationship in the Neotropics to include this kind of evidence in future research agendas, in order to generate new data on these particular human-animal relationships in the past. This is particularly needed since a number of marine and riverine mammal species are nowadays in danger and, hence, the zooarchaeological record is one of the few lines of evidence that can offer a long time perspective on the natural history of these species, as we discuss below.

A second aspect, closely linked to the previous one, that also might help to explain the heterogenity mentioned above has to do with the lines of evidence studied in each Neotropical area. For instance, the detailed historical data, many written at times of the Spanish conquest by chroniclers and missionaries, helps enormously to build research hypothesis in Mesoamerica and the Caribbean and is furthermore employed to contrast the zooarchaeological data with broader accounts. On the contrary, the current zooarchaeological research in the Southern South American region strongly rests on zooarchaeological data, both as a consequence of a research tradition that rely on independent explanation and a scarcer historical data avalailability as well. The differences in the ways regional archaeological explanations are constructed also points towards a third cause of heterogeneity: the proper condition of the regional zooarchaeological record of aquatic mammals in the Neotropics. The latter can be abundant in the South (Patagonia) and some northern areas (Baja California) but scarce in Mesoamerica and central Neotropical areas. Faunal resources in general are differentially abundant and diverse throughout the American continent.

As previously said (see introduction), the southern tip of South America is the southernmost emerged land of the austral hemisphere, except Antarctica, a continent only peopled recently by humans. More precisely it consists of a peninsula penetrating an oceanic hemisphere (Morello 1984; Iriondo 1999) and display a relatively poor insular fauna (Franklin 1982; Redford & Eisenberg 1992). This has impinged upon the human - animal relationship along the Holocene (see Muñoz and Mondini 2008, for a discussion). On the other hand, Mesoamerica and the tropical portion of northern South America have one of the mayor biodiversities of the world (SRE 2013). For example, Colombia is one of the countries with major biodiversity in the world, due to the strategic geographical location, which encompasses a variety of ecosistems. According to Trujillo et al. (2013), 10% of the fauna and flora word´s species is homed in this region. Such abundances of faunas can define the selection of only suitable (and easy accessible) taxa by aboriginal groups, explaining thus why other taxa (e.g. heavy aquatic animals) were not or only randomly hunted. The zooarchaeological data from Mesoamerica, one of the pivotal regions of the American continent where agriculture gained fundamental importance for the development of the Indian societies since early times, seem to indicate that the faunal resources obtainable near the settlements, in and around the cultivated fields formed by the milpa system were abundant enough to satisfy the needs of the people (see Götz 2008, 2014 for a discussion). In this sense, the tropical and subtropical areas of the American continent in general, and Mesoamerica in particular, are not only distinct to the other areas discussed in the book in terms of much less aquatic mammal species used by the aboriginal peoples, but especially in terms of a much higher abundance of other (presumably terrestrial) faunal taxa that was presumably exploited mainly during the milpa agricultural activities, which fulfilled the nutritional needs of the Indian population and which could be hunted without the technological necessities (heavy and solid boats, harpoons, nets and the like) and risks that requested the acquisition of aquatic mammals in the coastal waters.

The human-aquatic mammals relationships in time and space can also be analyzed from the substantive contribution the previous chapters clearly put forward. A review of the available evidence suggests that the use of species which inhabit shallow waters could be a diagnostic feature of aquatic mammal use in Mesoamerica and the Caribbean. This manly refers to the use of manatee and probably also to the West Indian seal, species that display a very different behavior and spatial use but which can be hunted at coastal settings and riversides without much technological effort and risk. The range of variation in the zooarchaeological signature that can be expected of the aquatic mammal use across time and space it is not yet well defined, but it is clear, after reviewing the chapters of this book, that different archaeological expectations on this issue can be linked to the ecological properties and subdivisions of the Neotropical region. A clear example in this sense is the Caribbean north cost of Colombia where the evidences for human consumption of aquatic mammals is restricted to manatee and even in this case the evidence is very scarce and seems to represent occasional captures. In contrast, other kind of aquatic taxa, reptiles and fishes mainly, have a very high representationas showed by recent regional zooarcheological data (see for example Ramos 2013; Ramos & Archila 2008; Oyuela, 1996).

In the insular environments of the southern tip of the continent, the use of canoes implied an active use of the sea channels, instread. The use of those resources can be explained as a result of an increase of the insularity gradient that the continent displays at the southernmost latitudes, involving an active use of the marine landscapes. At the opposite corner of the area covered by the book, Baja California, something similar can be suggested regarding the intense use of aquatic mammals by the inhabitants of the southern peninsula during late Holocene time.

In other words, the importance of riverine- water resources in the central and Neotropical areas and insularity related strategies in the northern and southernmost areas of the Neotropics can be discussed in terms of range of variations of human responses to the specific conditions of the environments where they evolved.

The relationships of human to other animals in their communities from a long-term perspective and broad spatial expression involve the rates at which tactics, actions, and decisions are repeated (Stiner 1993). Putting together this variabilitiy at a continental – e.g., Neotropical - scale is, as can be seen in the preceding chapters, a productive way to address the role of humans in communities mostly lacking other large social predators. As Muñoz and Mondini (2008) highlight, the ecological and historical properties at a continental scale may help to understand some of the patterns displayed in the archaeological record, such as those of South America, for instance the importance of an early use of aquatic resources not necessarily related to elevated human population densities, as some of the oldest instances of maritime adaptations in coastal Peru illustrate.

As a final result, the review offered in this book permits us to consider the different ways humans impacted on aquatic mammal populations in the past, in both prehistoric and historic times. Archeological data contribute, this way, to the knowledge of the natural history of the Neotropical aquatic mammals and help to evaluate the ecological status of aquatic mammals in different areas of this region.

Particularly, zooarchaeology has proven to be highly relevant to modern-day issues, and not strictly restricted to subjects regarding a distant past. That is why, in recent years, we have seen an increase in the number of publications devoted to discuss the results of zooarchaeological investigations applied to conservation problems and to explore on the nature of contributions that can be offered from this discipline in the conservation of current species and environments. Two good examples for these efforts are the works whose results are compiled in the books "Zooarchaeology and Conservation Biology", by Lyman and Cannon (2004) and "The Future from the Past", by Lauwerier and Plug (2004), both of which contain an important number of zooarchaeological researches in the Americas and other parts of the world strictly dedicated to these issues.

A series of important contributions that zooarqueological perspective can make to conservation biology ar highlighted by Lyman and Cannon (2004). It is argued that one of zooarchaeology's biggest contributions is its temporal perspective, since its range is more inclusive than the one used by most scientific fields and this allows designing more effective and broader techniques to study human activity's impacts temporal perspective, for example, has shown the importance of taking into account prehistoric ranges of the species that once lived on locations that are currently threatened and that contain endangered biota that scientists wish to recover. Thus, zooarchaeology has proven to be extremely useful in various ways. In the first place, it has stressed that human impact is not exclusive for the last thousand years, because it has happened since early moments of the human evolution process. This helps to dismantle the strong idea of "pristine" or "natural" environments, in the Americas especially, which has preconfigured many conservation strategies and constitues an influential concept in common imaginary. Also, it allows to deconstruct old categories and to propose new ones (like "historic landscape") that are more accurate and appropriate for the purposes of the kinds of studies. It also highlights that most conservation work assume biotas as intact units when responding to environmental change, which zooarchaeology has helped to prove wrong. Thus, as Lyman and Cannon mentioned (2004:22) "Such taxonomically individualistic responses to environmental change imply that the biological preserves of today are artifacts of the time when they were identified and created".

As mentioned above, zooarchaeology allowed studying phenomena that are difficult to monitor in a scientist's life span, which becomes important when considering that a specific landscape does not exists exclusively in a current time and in space, because its history is highly important as well. This specific history of a certain place is what allows zooarchaeology to be very useful in studying and helping to design strategies for particular places, taking into account the particularities of its geography and species. In general terms, then, it is clear that zooarchaeology has a great importance in the task of conservation, because it is the best tool at hand to use history as the best guide to take the most efficient and beneficial decisions today. At this point, it is especially important to settle some "reference conditions" to know precisely what can be done and can be expected from the recovery and conservation of a particular site, and zooarchaeology results appropiated for this.

Of course, regarding the scarceness of information about the use of aquatic mammals in the majority of areas in Mesoamerica, it results difficult to assess clear impact patterns that could help to determine how our results could improve actual policies of faunal conservation in the area. The Maya, for example, apparently did not impact significantly the aquatic mammal faunas, since both manatees and seals continued to exist in the area well beyond the Spanish Conquest. As Hairr (2011) states, the West Indian seal did not vanish before the 20th century, thus its disappearance was not due to a prehispanic use and impact neither to the landscape nor the faunal groups, but due to historic and modern exploitation of both, the coastal habitat and the animals.

A similar development can be seen in the recent and prehispanic history of the manatee. These slow and calm aquatic mammals were apparently seldom hunted, perhaps because the exploitation of the sea by the Mayas yielded enough food in form of mollusks and fish, plus occasional birds (see Götz 2012). Yet, Aridamerica indicated a different stage of human exploitation of aquatic mammals; in Baja California, sea mammals were readily hunted and consumed, since archaic times, a pattern quite similar to cases from Southern America discussed in our book. It is thus very probably the richness or scarceness of the environment and the availability of menu resources that determined if aquatic mammals were hunted; coastal areas where the general menu resources for pre-Clombian humans were scarce, saw more likely the development of cultural traits specialized in hunting sea mammals.

Those patterns indicate that even Indian groups could have sometimes decimate local sea mammal populations. An important difference between the local extinctions caused by Indian groups and the wide scale disappearance of marine mammals -such as the West Indian seal- provoked by modern humans may lie in the fact that Indian groups, such as the ones described by Zangrando, were limited to a relatively small area, while modern (post-Conquest) European societies spread over almost all different environments and areas, at least in order to hunt and provide food for the growing settlements in the New World.

What it is then that can be learned through this book? How can we use the archaeological information to support modern intends to conserve and protect the few aquatic mammals that are still left on this part of the globe? The analysis of prehispanic pattern shows that the exploitation was directed towards the faunal groups immediately needed and that were exploitable. But it shows on the other hand that those exploitations did not cause the complete extinction, but rather a level of adaptation between human cultures and aquatic mammal groups. What did in fact end or threaten the existence of some of the animals we treat in this book, such as the West Indian seal or the manatee, was not the pre-Colombian exploitation, but the modern hunt and more so, the occupation and perturbation of the refugia of those taxa.

REFERENCES

Franklin, W. L. (1982). Biology, ecology, and relationship to man of the South American camelids. In Mares, M. & H. Genoways (Eds). *Mammalian biology in South America, Lab. of Ecology, Special Publication* 7 (pp. 457-489). Pittsburgh, USA: The University of Pittsburg.

Götz, C. M., (2008). Coastal and Inland Patterns of Faunal Exploitation in the Prehispanic Northern Maya Lowlands. *Quaternary Internationa, 191*:154–169.

Götz, C. M. (2012). Caza y pesca prehispánicas en la costa norte peninsular yucateca. *Ancient Mesoamerica 23(2)*, 421-439.

Götz, C. M. (2014). La alimentación de los mayas prehispánicos vista desde la zooarqueología. *Anales de Antropología 48 (I)*, 167-199.

Hairr, J. (2011). *Caribbean monk seals - Lost seals of the Gulf of Mexico and Caribbean sea*. Landisville, USA: Coachwhip Publications.

Iriondo, M. (1999). Climatic changes in South American plains: records of a continent-scale oscillation. *Quaternary International 57/58*, 93-112.

Lauwerier R. & Plug, I. (2004) (Eds.). *The future from the past. Archaezoology in wildlife conservation and heritage management*. Chippenham, UK: Oxbow Books.

Lyman, R., & Cannon, K. (Eds.) (2004). *Zooarchaeology and Conservation Biology*. Salt Lake City, USA: The University of Utah Press.

Madriñán, S. & Sánchez, J.A. (Eds.) (2012). *Biodiversidad, Conservación y Desarrollo*. Bogotá, Colombia: Universidad de los Andes.

Morello, J. (1984). *Perfil ecológico de sudamérica. Características Estructurales de Sudamérica y su Relación con Espacios Relevantes del Planeta vol. 1*. Barcelona, España: Ediciones Cultura Hispánica, Instituto de Cooperacion Iberoamericana.

Muñoz, S. & Mondini, M. (2008). Long term human/animal interactions and their implications for hunter-gatherer archaeology in South America. In Papagianni, D., R. Layton & H. D. G., Maschner, (Eds.). *Time and Change: Archaeological and Anthropological Perspectives on the Long Term* (pp. 55-76). Oxford, UK: Oxbow Books.

National Geographic (1996). *La contaminación marina*. Available from: http://www. nationalgeographic.es/el-oceano/cuestiones-criticas-sobre-el-problemas-de-la-contaminacion-marina/cuestiones-criticas-sobre-el-problemas-de-la-contaminacion-marina].

Oyuela, R. (1996). The Study of collector Variability in the Transition to Sedebtary Food Producers in Northern Colombia. *Journal of World Prehistory 10(1)*, 49-93.

Ramos E. (2013). Crawling and walking at the time: challenges of the archaeology of the animals in the North of South America". In Götz, C. M. & K. F. Emery (Eds.). *The Archaeology of Mesoamerican Animals La Arqueología de los Animales de Mesoamérica* (pp. 531-555). Atlanta, USA: Lockwood Press.

Ramos, E. & Archila S. (2008). *Arqueología y Subsistencia en Tubará. Siglos IX-XVI*. Bogotá, Colombia: Centro de Estudios Socioculturales e Internacionales.

Redford, K. H. & Eisenberg, J. F. (1992). *Mammals of the Neotropics, The Southern Cone vol. 2*. Chicago, USA, Chicago University Press.

Secretaría de Relaciones Exteriores (SRE), (2013). Available at: http://embamex. sre.gob.mx/suecia/images/pdf/megadiversidadesp.pdf

Stiner, M. C. (1993). The place of hominids among predators: interspecific comparisons of food procurement and transport. In Hudson, J. (Ed.). *From Bones to Behavior. Ethnoarchaeological and Experimental Contributions to the Interpretation of Faunal Remains, Occassional Paper No. 21* (pp. 38-61). Carbondale, USA: Center for Archaeological Investigations, Southern Illinois University at Carbondale.

Trujillo, F., Gartner, A., Caicedo, D. & Diazgranados, M. C. (2013). *Diagnóstico del estado de conocimiento y conservación de los mamíferos acuáticos en Colombia*. Bogotá, Colombia: Ministerio de Ambiente y Desarrollo Sostenible, Fundación Omacha, Conservación Internacional y WWF.

EDITORS' CONTACT INFORMATION

Dr. A. Sebastián Muñoz
Investigador Adjunto
Profesor Adjunto
Consejo Nacional de Investigaciones Científicas y Técnicas
Laboratorio de Zooarqueología y Tafonomía de Zonas Áridas
IDACOR-CONICET, Museo de Antropología
Universidad Nacional de Córdoba
Av. H. Yrigoyen 174 (5000) Córdoba
Argentina
Tel: (0351) 4332105
E-mail: smunoz@conicet.gov.ar

Dr. Christopher Markus Götz
Profesor-Investigador Titular de tiempo completo
Responsable de Taller de Zooarqueología
Facultad de Ciencias Antropológicas
Universidad Autónoma de Yucatán
Km. 1 Carretera Mérida, Yucatán
México
Tel: 01 999 930 00 90 ext. 2007
E-mail: cgotz@uady.mx

Dr. Elizabeth Ramos Roca
Profesora Asociada
Departamento de Antropología
Universidad de los Andes
Cra la # 18A-10 Edificio Franco oficina GB624
Bogotá- Colombia
Tel: 57-1-3394949 ext. 2555
E-mail: eramosroca@uniandes.edu.co

INDEX

C

D

E

F

G

H

M

N

O

P

Q

R

T